AF324897

Toward Sustainable and Economic Smart Mobility

Shaping the Future of Smart Cities

Toward Sustainable and Economic Smart Mobility

Shaping the Future of Smart Cities

Editors

Max Eiza
University of Central Lancashire, UK

Yue Cao
Lancaster University, UK

Lexi Xu
China Unicom, China &
Beijing University of Posts and Telecommunications, China

NEW JERSEY · LONDON · SINGAPORE · BEIJING · SHANGHAI · HONG KONG · TAIPEI · CHENNAI · TOKYO

Published by

World Scientific Publishing Europe Ltd.

57 Shelton Street, Covent Garden, London WC2H 9HE

Head office: 5 Toh Tuck Link, Singapore 596224

USA office: 27 Warren Street, Suite 401-402, Hackensack, NJ 07601

Library of Congress Cataloging-in-Publication Data

Names: Eiza, Max, editor. | Cao, Yue, editor. | Xu, Lexi, editor.
Title: Toward sustainable and economic smart mobility : shaping the future of smart cities /
    editors, Max Eiza, University of Central Lancashire, UK, Yue Cao, Lancaster University, UK,
    Lexi Xu, China Unicom, China & Beijing University of Posts and Telecommunications, China.
Description: Hackensack, NJ : World Scientific, 2020. |
    Includes bibliographical references and index.
Identifiers: LCCN 2019047993 | ISBN 9781786347855 (hardcover) | ISBN 9781786347862 (ebook) |
    ISBN 9781786347879 (ebook other)
Subjects: LCSH: Intelligent transportation systems. | Transportation--Technological innovations. |
    Transportation--Management. | Communication and traffic. | Sustainable development. |
    City planning--Technological innovations.
Classification: LCC TE228.3 .T69 2020 | DDC 388.3/12--dc23
LC record available at https://lccn.loc.gov/2019047993

British Library Cataloguing-in-Publication Data
A catalogue record for this book is available from the British Library.

For any available supplementary material, please visit
https://www.worldscientific.com/worldscibooks/10.1142/Q0233#t=suppl

Desk Editors: Anthony Alexander/Jennifer Brough/Shi Ying Koe

Typeset by Stallion Press
Email: enquiries@stallionpress.com

Preface

Have a look around and tell us how many smart devices you can see? Whether it is your smart mobile phone, smart watch, smart car, smart building, or smart supermarket, you are surrounded by devices that interact with and collect information about you and your environment. Soon we will transition from living in this smart environment to living in what we call a smart city, and this will improve many aspects of our lives and the quality of services (QoS) we receive. One of the main improvements these smart cities are aiming to achieve is better, greener and more efficient transportation, also known as smart mobility. The recent developments in those smart devices, the phenomenon of the internet of things (IoT) and vehicular communications are revolutionizing the future of smart cities and smart transportation. With the rapid integration of those smart devices, we are heading to a new era of highly connected and environmentally friendly ecosystems in which we will thrive.

This book brings the element of smart mobility into the spotlight, including its latest developments and applications in the context of smart cities, ranging from the latest projects that have been implemented in the world to emerging technologies, such as Big Data and artificial intelligence (AI), research directions, security, user privacy and future trends. It also investigates how smart mobility can facilitate future transportation systems in smart cities. This is evident in the main theme of the book, which focuses on smart mobility services for smart cities, such as smart traffic control systems and autonomous valet parking. The book sheds light on how smart mobility can be used for the provision of better services in a highly connected heterogeneous network such as a smart city.

Throughout this book, we aim to provide a unique overview of the broad developments in smart mobility, in the context of smart cities, from different technical perspectives. The book gives a broad overview and deep insights into the latest developments in smart mobility, while offering a unique opportunity for the readers to explore different perspectives and ideas for future research, as well as to have more of their questions answered before bringing smart cities to life. Moreover, it provides a reference for professionals and researchers in the areas of smart mobility (e.g., autonomous valet parking, passenger trajectory data, smart traffic control systems) and recent technical trends in their enabling technologies.

This is the first work that provides such insights into the latest developments in those areas as most of the related works are either very narrow or outdated. The book is a valuable resource for researchers and professionals working in smart environment research and the automotive industry. Notable, talented and internationally recognized authors from academia and industry have contributed to this book, covering an array of timely, topical and emerging topics in smart mobility. These contributions fall into the four main themes in this book:

(1) smart mobility trends and challenges;
(2) smart mobility services for smart cities;
(3) smart mobility for optimizing data networks in smart cities;
(4) smart mobility security and the user's privacy.

Organization of Chapters

This book is organized into nine chapters, and each chapter is written by one or more topical area experts. The chapters are arranged in an order that makes them easy to read and follow to help the reader gain the most from this book.

Chapter 1 introduces projects in smart cities around the world and the latest trends and challenges. Chapter 2 illustrates the application of Big Data and AI technologies in smart mobility and discusses the workflow of adopting such technologies in smart cities.

Chapter 3 discusses the future smart traffic control systems based on the latest vehicular communication technologies, and how to overcome the shortcomings of the current traffic control systems. Chapter 4 shows how mobile phones can be used as representatives of vehicles on the road to improve automatic monitoring of the road network and provide accurate and real-time road information. Chapter 5 reviews the recent modern data-driven models for passenger trajectory generation on a public transport network and discusses the benefits of having such data while highlighting the gaps in knowledge that still need to be filled. In Chapter 6, the authors propose a novel parking schedule scheme for long-range autonomous vehicle parking that considers the mobility of these vehicles, their fuel consumption and their journey time.

In Chapter 7, a brief summary on the state-of-the-art mobile data offloading for smart mobility in heterogenous networks is presented. The two main types of networks are discussed: infrastructure-based networks and infrastructureless networks. Then, the authors have identified the key challenges that need to be addressed in this field. Chapter 8 develops a hybrid intelligence model for load balancing in smart environments and evaluates its performance via simulation experiments. Finally, Chapter 9 deals with the user's security and privacy issues in the context of smart cities by introducing UMBRELLA: a novel user demand privacy-preserving framework that strikes a balance between users and attackers in terms of the privacy demanded.

About the Editors

 **Max Hashem Eiza** received his BSc (Hons) in Computer Networks and Operating Systems from Damascus University in 2007. In 2010, he received his MSc in Data Communication Systems, with Distinction, from Brunel University London. He also received a PhD (specializing in the Secure QoS Routing Algorithm for Vehicular Networks) from Brunel University London in 2015. From 2010 to 2015, Max was a research student at Brunel University London, UK. During 2015–2016, he worked as a Research Assistant (cybersecurity) at Liverpool John Moores University, UK. From 2016 to the present date, he has worked as a Lecturer in the Computing Department, teaching Computer Networks and Security at the University of Central Lancashire, UK. Max has also actively participated in various external professional activities. For instance, from 2013 to the present date, he has worked as a regular reviewer for journals such as *IEEE Transactions on Vehicular Technology*, *Wiley Security and Communication Networks*, and *MDPI Sensors*. In 2015, he was the TPC member of *IEEE GLOBECOM* 2015. In 2016, he was the TPC member of *IEEE CCNC* 2016, and was also the editor of the *International Journal of Vehicle Information and Communication Systems* (*IJVICS*). In 2017, Max was the TPC member of *IEEE CCNC* 2017 and *IEEE GLOBECOM* 2017, and was the editor of the *International Journal of Cyber Situational Awareness* (*IJCSA*). Apart from this, he has presented and published several papers in prestigious IEEE conferences and journals, and has served in the committees of several prestigious IEEE conferences and workshops, such as *IEEE GLOBECOM*

and *IEEE CCNC*. He has authored publications and presentations in various journals, such as *IEEE Vehicular Technology Magazine*, and *IEEE Transactions on Vehicular Technology*, and contributed to the Proceedings of the *IEEE International Conference on Communications (ICC)*, 2016, Kuala Lumpur, Malaysia. He has also co-authored a chapter in the book *Vehicular Social Networks*. Max's research is centered on the increasingly important theme of distributed network security and data privacy with the aim of developing novel schemes/protocols for various applications. Specifically, his research interests encompass the emerging security, privacy, trust, and identity management issues in the following areas: vehicular networks, the internet of things, Big Data, cloud computing, smart grid, and smart city. Currently, Max is involved in the EU ERDF project "Digital First", enabling SMEs to embrace digital technologies and raise awareness about cyber security.

Yue Cao received his PhD from the Institute for Communication Systems (ICS), University of Surrey, UK in 2013. He was a Research Fellow at ICS, University of Surrey, UK; a Lecturer and Senior Lecturer at the Department of Computer and Information Sciences, Northumbria University, UK; and International Lecturer at the School of Computing and Communications, Lancaster University, UK. He is the Associate Editor of *IEEE Access*, Springer's *EURASIP Journal on Wireless Communications and Networking*, *KSII Transactions on Internet and Information Systems*, and IGI Global's *International Journal of Vehicular Telematics and Infotainment Systems*.

Lexi Xu received his PhD from the School of Electronic Engineering and Computer Science, Queen Mary University of London, London, UK in 2013. Xu is currently a Senior Engineer at the Network Technology Research Institute, China United Network Communications Corporation (China Unicom), as well as a senior researcher at the National Engineering Laboratory for Next Generation Internet Broadband Service Application. He has been actively involved in 30 Big Data research and network optimization projects. He has published more than 70 technical papers and two international conference proceedings, and he has applied for more than 20 patents. He is the Industrial Tutor of BUPT, NHU, and SAU. He is also a China Unicom delegate in ITU, ETSI, and CCSA. He served as the Workshop Chair/TPC Co-Chair of the *ICSINC* (2015–2019), *IEEE ISCIT* (2016), *5GWN* (2017), and *IEEE IUCC* (2019) workshops. His research interests include Big Data, self-organizing networks, smart mobility, and radio resource management in LTE-A/5G.

Acknowledgments

We would like to thank everyone who contributed to the success of this book including the authors who worked very hard to produce high-quality content on the topics of smart mobility and smart cities.

We also would like to thank the publishing and marketing teams at World Scientific Publishing, in particular Jennifer Brough and Catharina Weijman who worked with us on this book project from the beginning.

Contents

Smart Mobility:
Challenges and Trends

Ruizhi Liao

The Chinese University of Hong Kong, Shenzhen,
Guangdong, 518172, P. R. China
rzliao@cuhk.edu.cn

1 Introduction

According to the INRIX Global Traffic Scorecard [1], Los Angeles topped the list of the world's most congested cities for the sixth year in a row. The report shows that drivers in Los Angeles had to sit for 102 hours in congestion during peak times in 2017, which, as calculated by INRIX, cost each driver $2,828 and the city over $19 billion. Metropolises around the world share similar traffic pains with Los Angeles, which is followed by Moscow and New York (both at 91 hours), Sao Paulo (86 hours), San Francisco (79 hours), Bogota (75 hours) and London (74 hours).

Dynamic pricing to control traffic and adding new road infrastructure, among others, are popular measures to relieve the city of congestion. These traditional ways are effective, but they come with notable drawbacks such as the following:

- In 2003, London set up a congestion zone to restrict vehicles entering into the central area. A vehicle entering into the central area would be charged a fixed fee of £11.5 a day [7]. One year later, the number of cars entering into the area dropped by 18%, and the traffic delay was reduced by 30%. However, these pretty improvements were met with angry commuters and worried business leaders who feared the increased transportation cost may frustrate industries.

- Regarding the addition of new infrastructure, it may involve environmental assessment, land acquirement, the endorsements of city councils, the relocation of residents, construction and maintenance. The main problems with the measure are speed and cost.

Clearly, new mobility modes are desired. In the past decade, smart mobility solutions have started to emerge against the traditional ones [15], e.g., assisted/self-driving, intelligent logistics, smart parking and sharing economy (ride-/car-/bike-sharing). These innovative mobility solutions not only facilitate quicker movement but also alleviate urban pains, such as traffic congestion, oil waste and pollution.

This chapter explores the new path toward smart mobility. It first provides an overview of high-profile smart mobility/smart cities projects by identifying their data sources and enabling technologies; then, it raises the key issues on technical, legal and social challenges that we have to address before a smart city/mobility project can be called a success.

2 Overview of Smart Mobility Projects

Self-driving is still in its infancy, but ride-/car-/bike-sharing and smart parking have already been constructing their respective mobility ecosystems for a while. In the past, we relied on our cars or public transportation to travel around. Now, in metropolises, the way we get to our destination is changing. These changes indicate that new ways of transportation are emerging, e.g., car ownership is no longer pursued, and customized transportation based on crowdsourced data is becoming the trend [18].

This emerging trend of personal mobility is called mobility as a service (MaaS), a kind of user-tailored mobility service. It benefits from the information and communication innovations that synthesize all aspects of transportation, e.g., trip planning, booking, parking, payment and ticketing. In order to meet users' needs, MaaS provides the most suitable transportation means, be it public buses, taxis, rental cars or even ride-/car-/bike-sharing. Thus, MaaS means the most convenient solutions for users.

In this section, we sample some representative MaaS projects. In order to give readers a very quick reference, the main features of these projects are summarized in Table 1, while more detailed descriptions are presented in Section 2.1.

Table 1. Summary of reviewed MaaS projects.

Projects (year)	Related services	Portal	Actors	Technologies
SUNSET (2011)	Sustainable mobility solutions	App	Universities, tech companies	Social networks, incentive algorithms
Sentilo (2012)	Smart parking/ transit	Web	City hall, tech companies, nearby cities	Open-source platform, sensors
Helsinki 2025 Vision (2014)	On demand, door-to-door mobility	App	City hall, tech companies	Big Data, cloud computing
Smart Mobility 2030 (2014)	Crowdsourced, door-to-door mobility	App	Transport authority	Big Data, cloud computing
Smart Columbus (2017)	Smart mobility, smart parking	App, web	City hall, tech companies	Big Data, cloud computing, sensors

2.1 *Overview of metropolitan MaaS projects*

The reviewed projects are divided into groups by how they derive their key data for mobility planning. As each individual project often ingests data from multiple places, we will categorize them based on the main sources of their data.

2.1.1 *Data source: Social networks*

In 2011, an EU project called "Sustainable Social Networking Services for Transport (SUNSET)" was created [9]. The objectives of SUNSET are to reduce congestion, cut down pollution and improve safety by encouraging users to change their mobility patterns. The widespread use of smart phones and ubiquitous internet access are assumed. Users regularly share their mobility means on social networks. By analyzing those patterns and the traffic conditions monitored by roadside sensors, SUNSET algorithms will list up several transport solutions. In order to attract more users, SUNSET devised interesting and rewarding incentives to stimulate users to adopt the congestion-free and environment-friendly routes. For example, users share their mobility profiles on social networks, then they will receive useful route suggestions or rewards if they adopt the suggested ones. More importantly, by having residents' mobility preferences, the traffic administration is able to fine-tune traffic polices, redesign road infrastructure or offer more compelling incentives. It is claimed to be a win–win situation for all stakeholders of cities.

As a follow-up, a smartphone app called Tripzoom [6] was developed as a proof of concept to facilitate the sharing of traveling information. Also, inspired by SUNSET, the startup Mobidot (www.mobidot.nl) was created. Another research project, EMPOWER [17] has a similar idea to SUNSET: ingesting mobility profiles on social networks and guiding users to adopt sustainable solutions by offering rewards.

2.1.2 *Data source: Open data*

Apart from exploring information on social networks, traffic statistics from thousands of sensors across cities are another important data source for mobility innovations. In 2012, Barcelona initiated an open-source sensor data processing platform called Sentilo [4], which is designed for cities that look for openness and easy interoperability. It collects data from various sources and underpins the deployment of smart parking and smart transit services. Sentilo was built with the aim of openness; thus, the source codes were shared under an open-source license, which is available at www. sentilo.io. For that reason, many surrounding cities (e.g., Terrasa, Reus and Cambrils) and their municipal services are utilizing the platform. Fastprk, which deployed 500 parking sensors in Barcelona, is one of the many applications based on Sentilo that provide smart mobility services [12].

Nowadays, metropolises usually organize and release their cities' data on a regular basis, e.g., widespread sensor data, demographic data and industrial data, in the Open Data format, e.g., London DATASTORE (https://data.london.gov.uk/dataset) and the NYC Open Data website (https://opendata.cityofnewyork.us). These Open Data platforms are treasuries driving innovative applications, e.g., using road monitoring sensor data to recognize traffic patterns [16]. A recent Deloitte report indicated that, benefiting from the Open Data released by Transport for London, more than 750 apps were created. These apps are actively used by 42% of Londoners, contributing to London's economy by £130 million each year [3].

Although most Open Data across the globe are free to access, there are exceptions. Copenhagen built the world's first city data marketplace in May 2016, where public and private data are traded for a price.

2.1.3 *Data source: Users' demands*

Helsinki is one of the leading cities to initiate a MaaS project [13]. In 2012, Kutsuplus, a pioneering on-demand minibus service, was launched in the capital of Finland. The most important service provided by Kutsuplus was to group users traveling in similar directions into minibuses. Kutsuplus was a sort of hybrid that cost the price of a bus ride but provided the convenience of a taxi. Although Kutsuplus was discontinued after 2 years due to lack of density and scale, it was the world's first attempt to reinvent the concept of car-sharing [8].

In 2014, Helsinki announced an aspiring plan that the city will eliminate private car ownership by 2025. It is going to be so convenient for users to travel around the city that they will naturally abandon personal cars, not because they are forced to but because the alternative mobility modes are just more appealing [10]. The design of today's transportation systems is infrastructure–vehicle-centered, focusing on roads, subways, buses and cars. In order to achieve the ambitious Helsinki 2025 vision, a rethinking of this overall design is required.

Helsinki put forward a very different model by shifting the design focus from the infrastructure and vehicles to users: how to transport each traveler from one place to another quickly and efficiently. The Helsinki 2025 vision relies on an on-demand mobility system that combines all sorts of transportation tools and intelligently adopts the best way to get to where users want to go. In 2016, an app named "Whim" was created to enable Helsinki residents to plan and pay for all modes of transportation within the app. If no single mode can cover the entire door-to-door trip, a combination of bus, taxi, or ride-/car-/bike-sharing [5] will be suggested. The features of the sharing mobility are summarized as follows:

(1) **Ride-sharing** utilizes the empty car seats bound for the same direction. It does not add any new vehicles to the traffic.
(2) **Car-sharing** is a model where people rent cars for short periods of time. It is attractive to users who occasionally need vehicles.
(3) **Bike-sharing** has been on the upswing in recent years. It is a good mobility tool for short-distance commuting, e.g., first-/last-mile connections.

In 2014, Singapore launched the Smart Mobility 2030 Strategic Plan [11], aiming to build a more connected and interactive community. Self-Driving for future mobility, Open Data for urban transportation and real-time information for demand-driven transportation were actively discussed. In 2015, Singapore piloted its first crowdsourced bus services called "Beeline" to explore innovative solutions in first-/last-mile connections. Beeline is a Big Data- and cloud-based smart mobility platform. Users can not only book a seat through the app but also, if they find no suitable existing routes, are able to crowdstart new routes. For example, if many people live in proximity and travel to work in the same industrial area, they input their respective desired pick-up/drop-off locations, then the system recommends some existing routes. If there is no suitable route, users' demands are recorded and analyzed to intelligently decide where the new routes will be. Similar services are also available in other metropolises across the globe, such as Chariot in San Francisco, Shuttl in Delhi, and DiDi Bus in Beijing and Shenzhen.

In January 2017, the U.S. Department of Transportation announced that the City of Columbus, Ohio, was the winner of the first-ever smart city challenge, in which around 80 U.S. cities competed for funding to develop intelligent transportation systems and utilize digital technology to upgrade urban mobility. On securing the funding, the city of Columbus rolled out nine interconnected projects to execute its ambitious vision of a smart city. A brief overview of the nine projects is given as follows [2]:

(1) **Smart Columbus Operating System** is a web-based platform that serves as a data center which digests the collected data and distributes data to other applications.

(2) **Connected Vehicle Environment** installs sensors on vehicles and at street intersections to set up vehicle-to-infrastructure networks, aiming to avoid collisions and ingest traffic data.

(3) **Multimodal Trip Planning and Common Payment System** is the user interface that allows one to plan and pay for seamless trips, which may involve multiple service and parking providers.

(4) **Smart Mobility Hubs** are places linked to transportation, including Wi-Fi access points, trip-planning booths and pick-up/drop-off locations for multi-mode transportation.

(5) **Mobility Assistance** is an app that facilitates people with disabilities. The app aims to allow this group of residents to navigate public transportation safely and independently.

(6) **Prenatal Trip Assistance** is an app to help expectant mothers on Medicaid to get to non-emergency medical checks with flexible and reliable two-way transportation.

(7) **Event Parking Management** integrates multiple parking providers into a single parking and reservation services application, with which the direct parking routes in large events will reduce congestion significantly.

(8) **Connected Electric Autonomous Vehicles** will be deployed in the business and retail area of Easton to provide the first-/last-mile connections between a business center and bus stops.

(9) **Truck Platooning** shows Columbus's efforts to provide intelligent logistics for freight companies. It aims to connect and coordinate trucks' actions with one another, e.g., speeds and braking, to save fuel, reduce emissions and improve freight mobility.

3 Challenges of Smart Mobility Projects

In this section, insights will be provided on technical challenges, legal challenges and social challenges, including financing and maintaining the smart mobility/smart cities projects, data ownership and data privacy.

3.1 *Thoughts on technical challenges*

Information security, payment security, data privacy and heterogeneous networks are classical challenges common in any information system and will not disappear in urban mobility systems. In this section, we would like to put forward the following challenges that are specific to smart mobility systems:

- **Management of Massive Networks:** Columbus has an ambitious plan which implies that massive interconnected networks, which will ubiquitously connect users, transport administrations, infrastructure (e.g., roads, lampposts, sensors and bridges) and transportation providers (e.g., car rental, public transportation and ride-/car-/bike-sharing services), are inevitable. Both cloud computing and fog/edge intelligence need to be developed, where the former can provide high-level business analytics and the latter can handle real-time applications.
- **Data Acquisition:** Three information sources were discussed in the previous section: social networks, Open Data (sensors and various statistics from city agencies) and users' demands.
  - It is difficult to quickly and directly extrapolate the useful information for mobility solutions via social networks, where natural language processing is needed.
  - The problem that Open Data faces is data quality. Many datasets turn out to be garbage that just adds more of a burden to the processing.

It is exactly why the premium and quality data may have a price, as discussed in the section on the Copenhagen data marketplace.

- User-tailored, on-demand or crowdsourced methods are the most efficient methods to estimate users' demands. The logic behind this is straightforward: each individual's request is satisfied with suitable mobility solutions. However, satisfying the spatially distributed users' requests with feasible solutions is very challenging.

- **Transportation Modeling:** The transportation modeling is important. A good transportation model can accurately estimate travel demands, offering valuable reference to policymakers and mobility planners so that they can better organize available resources. Machine learning (ML) algorithms that combine and analyze historical data, real-time sensor data, events or weather conditions can give accurate predictions. Unfortunately, we did not observe many discussions on transportation modeling in the sampled projects.

3.2 *Thoughts on legal challenges*

- Legal challenges arise because the new mobility ecosystem involves different players, e.g., payment companies, public–private transportation providers, city administration and users. It is no easy task to coordinate with these actors who are operating under different rules.
- The other legal challenge is privacy. A new mobility ecosystem means new data relationships. How much privacy does each actor need to give up to the system? What is the legal status of data? Is data an infrastructure? Is data a service? Is data a product? Why do these questions matter? Because the answers determine who owns the data and how to regulate who can access the data, and in what conditions the data can be accessed.

3.3 *Thoughts on social challenges*

- One of the biggest concerns is job losses. Similar to the logic that states that robots will replace human workers, the introduction of self-driving vehicles is causing truck and taxi drivers to worry as people will not require drivers any more. Governments need to make sure that retraining will be offered, and that new job opportunities will be created as the new mobility ecosystem becomes mature.
- The emergence of a new mobility ecosystem presents both huge opportunities and challenges to legacy transport players. They will face

increasingly stiff competition from the new mobility actors. If they do not adapt or are merged into the new ecosystem, they will be excluded from the major markets and eventually fade out.

- Financing is another big challenge. Most of the time, city administrations or national or international research agencies initiate the first move. Then, it goes to continuous development and maintenance. Who pays the bill? A cooperative arrangement involving public and private actors, namely, a public–private partnership, has worked well in the past. However, innovative models of collaboration, which can work within the increasingly diverse ecosystem of public, private and sometimes non-profit entities, are needed.
- The fourth challenge is related to the digital divide. Not all users are equally comfortable with using new mobility services. How can one really involve all citizens such that they benefit from the innovations?

4 Summary

There is no doubt that citizens will be the primary beneficiaries of the new mobility services. The city administrations face technical, legal and social challenges in ensuring that the complex integrated mobility ecosystem helps to achieve cities' common goals, such as higher efficiency, lower congestion, fewer traffic accidents, easier parking and better environment.

Cities such as Singapore, Helsinki and Barcelona are early models of what smart mobility will be like. Many private companies are also actively pursuing even deeper and more comprehensive integration. For example, in 2015, Ford collaborated with Georgia Tech to use sensors that come preinstalled on cars to search for roadside parking spaces [14]. In 2016, Ford acquired the startup Chariot to consolidate its long-term smart mobility strategy. In 2018, Ford launched Transportation Mobility Cloud, which aims to streamline information flow various components in the transportation ecosystem.

Smart mobility is in its nascent stages and still evolving, but the following observations suggest it might develop quickly and become mature in the near future:

- The common problems faced by cities, such as congestion and pollution, make them willing to adopt the sustainable mobility solutions.
- The smart mobility solutions are very appealing to city dwellers because of the problems they face when navigating their cities.

- The technologies that enable new mobility solutions are almost ready. Now, the most difficult part in making smart mobility a success would be integration.

Acknowledgment

This work was partially supported by the Shenzhen Science and Technology Innovation Committee with Grant Nos. JCYJ20170818103636337 and ZDSYS20170725140921348.

References

[1] Cookson, G., & Pishue, B. (2017). Inrix global traffic scorecard. INRIX Research.

[2] City of Columbus (2018). Concept of operations for the multimodal trip planning application/common payment system for the smart columbus demonstration program. White paper.

[3] Deloitte (2017). Assessing the value of TfL's open data and digital partnerships. http://content.tfl.gov.uk/deloitte-report-tfl-open-data.pdf [Accessed February 2019].

[4] Farrés, J. C. (2015). Barcelona noise monitoring network. In *Proceedings of the EuroNoise*, Barcelona, Spain, pp. 218–220.

[5] Goodall, W., Dovey, T., Bornstein, J., & Bonthron, B. (2017). The rise of mobility as a service. *Deloitte Rev*, 20, 112–129.

[6] Holleis, P., Luther, M., Broll, G., Cao, H., Koolwaaij, J., Peddemors, A., Ebben, P., Wibbels, M., Jacobs, K. & Raaphorst, S. (2012). TRIPZOOM: A system to motivate sustainable urban mobility. In *1st International Conference on Smart Systems, Devices and Technologies*, pp. 101–104.

[7] Ison, S., & Rye, T. (2005). Implementing road user charging: The lessons learnt from Hong Kong, Cambridge and Central London. *Transport Reviews*, 25(4), 451–465.

[8] Jokinen, J. P., Sihvola, T., & Mladenovic, M. N. (2017). Policy lessons from the flexible transport service pilot Kutsuplus in the Helsinki Capital Region. *Transport Policy*, 76, 123–133.

[9] Kusumastuti, D. (2012). *SUNSET. Sustainable Social Network Services for Transport. Deliverable D3.3: Impact of Incentives*. Enschede, The Netherlands: University of Twente.

[10] Kargas, C. (2015). SEAMless^{TM} Mobility: Shared, electric, autonomous, multimodal mobility. In *EIC Climate Change Technology Conference*, Montreal, Quebec, Canada.

[11] Land Transport Authority (2014). Smart mobility 2030 — ITS strategic plan for Singapore. White paper.

[12] Moskvitch, K. (2016). Barcelona: The world's Smart City? *Engineering and Technology*, 11(5), 48–51.

[13] Rissanen, K. (2016). Kutsuplus — Final Report. Helsinki: Helsinki Regional Transport Authority (HSL), 39.

[14] Roman, C., Liao, R., Ball, P., Ou, S., & de Heaver, M. (2018). Detecting on-street parking spaces in smart cities: Performance evaluation of fixed and mobile sensing systems. *IEEE Transactions on Intelligent Transportation Systems*, 19(7), 2234–2245.

[15] Schaffers, H., Komninos, N., Pallot, M., Trousse, B., Nilsson, M., & Oliveira, A. (2011). Smart cities and the future internet: Towards cooperation frameworks for open innovation. In *The Future Internet Assembly*. Springer, Berlin, Heidelberg, pp. 431–446.

[16] Soriano, F. R., Samper-Zapater, J. J., Martinez-Dura, J. J., Cirilo-Gimeno, R. V., & Plume, J. M. (2018). Smart mobility trends: Open data and other tools. *IEEE Intelligent Transportation Systems Magazine* 10(2), 6–16.

[17] van Amelsfort, D., & Hjalmarsson, A. (2016). Business models for incentive-based mobility services for changing traveller behaviour. In *23rd ITS World Congress*, 10–14 October 2016, Melbourne, Australia.

[18] Zhou, X., Wang, M., & Li, D. (2017). From stay to play — A travel planning tool based on crowdsourcing user-generated contents. *Applied Geography*, 78, 1–11.

Chapter 2

Big Data and Artificial Intelligence for Communications in Smart Mobility

Liang Zhao[*,§], Lexi Xu[†,¶] and Jiaxing Shang[‡,||]

*School of Computer Science, Shenyang Aerospace University, China
†China United Network Communications Corporation & Beijing University of Posts and Telecommunications, Beijing, China
‡College of Computer Science, Chongqing University, China
§lzhao@sau.edu.cn
¶xulx29@chinaunicom.cn
||shangjx@cqu.edu.cn

1 Introduction

Nowadays, the fast growth of the metropolitan area has been accompanied by a rapidly increasing number of vehicles as well as a rapidly increasing volume of road traffic which compares to that found in major cities worldwide. This also leads to problems such as an increase in traffic congestion, as well as other incidents that occur on highways as well as urban and rural roads. To address the above issues, vehicular communication is urgently needed to support the development of an intelligent transport system (ITS). Both academia and industry agree that vehicular networking and communication can extend the sensing level of vehicles to enable safe driving.

We can see that smart communication could integrate the computation power of vehicles, roadside infrastructures and cloud servers. Autonomous driving is the final goal of communication in the smart mobility scenario. It is necessary to realize several types of communication including vehicle-to-vehicle (V2V) communication, vehicle-to-infrastructure

(V2I) communication and vehicle-to-pedestrian (V2P) communication, to achieve this. All together, this is known as vehicle-to-everything (V2X) technology. Then the V2X technology can allow collaborations among vehicles, pedestrians and infrastructure to avoid crashes and reduce traffic jams. For example, it is hoped that V2X technology will save more than 25,000 lives per year in Europe.

We could denote vehicular communication as the top use-case for communications in smart mobility. Besides, technologies such as 4G/5G and wireless sensor networks (WSNs) could provide efficient communication that would result in them having a significant number of applications in smart mobility.

From this, we can see that the communication and networking paradigms play essential roles in enabling connectivity among different mobile devices (including vehicles and underwater autonomous vehicles (UAVs)) in smart cities. The sensory devices equipped in autonomous vehicles should be adapted to improve communication among vehicles, roadside units or even pedestrians should also be assured to ensure efficient and safe driving. There are three main benefits of harnessing communications to produce smart mobility. First, if we refer to sensors such as cameras, radar and LIDAR as the eyes of autonomous vehicles, then network communication is like the mouths and ears of human beings, which can be used as an alternate means of sensing the surroundings. Network communication allows for more accurate positioning, recognizing and information exchanging. Second, if network communication is used, the number of sensors needed per vehicle could be ascertained, which could lower the price of future autonomous vehicles and make them affordable. Third, as the computing power of the vehicle is limited, in particular, when it comes to real-time processing machine learning (ML) and terabyte-level data processing, we can shift the computation duty from the vehicle side to the edge, fog or even cloud with fast network communication, in order to save the computation resources in the car.

Due to the rapid development of vehicular communication, mobile communication and computer technology, the vehicular and mobile industry predict the challenge of data explosion. Big Data is characterized as "4-V", in which the 4 Vs stand for "Volume", "Variety", "Velocity" and "Veracity". Recently, the power of Big Data has been increasingly exploited in a wide range of applications, and it has been pushed to the forefront in both academia and industry. It is vital to investigate how to collect Big Data from such a sector and how to apply the data so as to innovate and enhance the quality of communication in the smart mobility scenario.

Meanwhile, artificial intelligence (AI) has achieved a significant breakthrough in providing high efficiency and adaptability in a multitude of fields, including computer vision, the automotive industry, financial analysis and so on. AI is a promising method for dealing with the dynamic and large-scale topology of computer networks and mobile networks, so we shall explore the applications of AI techniques, alongside Big Data and various ML algorithms, to smart mobility communications. These techniques include statistical learning, neural networks, deep learning and so on.

In this chapter, we mainly focus on introducing and presenting the existing applications of Big Data and AI as well as ML to communication paradigms for smart mobility. This chapter is organized as follows. In Section 2, the theory, system and usage of Big Data are presented. In Section 3, the basic approach to AI and ML is proposed in which the learning techniques, including supervised learning, unsupervised learning, semi-supervised learning and reinforcement learning (RL), are revealed. Then, in Section 4, the existing application of AI in all the layers of network communications for smart mobility is reviewed and discussed. Finally, in Section 5, we discuss the fundamental workflow of applying Big Data and AI in smart mobility communications and conclude.

2 Big Data for Smart Mobility

2.1 *What is Big Data?*

2.1.1 *The origin of Big Data*

Big Data has become a hot topic in both academia and industry in recent years due to progress in computer techniques and information. In 1997, Michael Cox and David Ellsworth published a groundbreaking article that used the term "Big Data" and then popularized the term and concept of Big Data in the 1990s.

Initially, "Big Data" indicated that data research and utilization are extremely complicated and big. Therefore, conventional data analysis tools and data processing software are not capable of addressing these data resources.

During 2000–2010, due to the fast development of mobile internet and smart terminals, many industries experienced data explosion. "Big Data" has changed people's lifestyles and their ways of thinking and working. In these circumstance, "Big Data" gradually began to include the employment of a series of algorithms, including data processing technologies, ML algorithms and prediction analysis algorithms, to deal with the data resources and extract valuable information from the data.

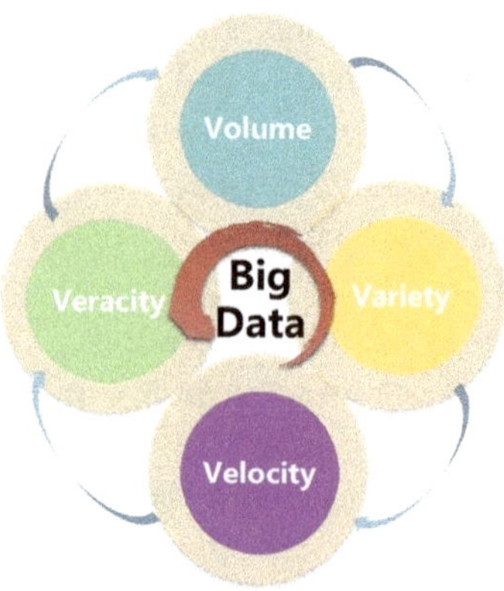

Fig. 1.   Big Data and 4V characteristics.

In 2016, an article gave the "3V" description of Big Data, namely, "Big Data represents the information assets characterized by such a high volume, velocity and variety as to require specific technology and analytical methods for its transformation into value" [1]. Furthermore, since the data quality directly impacts the value of Big Data, the term "veracity" is also employed to describe this feature of Big Data (see Fig. 1).

Recently, the above-mentioned "4V" has become a widely used method to describe "Big Data" in both academia and industry.

2.1.2   *The "4V" characteristics of Big Data*

Volume: Volume implies that the quantity of data is extremely large. The large volume of data always indicates that Big Data has a high value as it could yield potential insights when analyzed with the appropriate data mining algorithms and applications.

Variety: Variety implies that there are diverse data types and data features. There are diverse data sources in the data generation stage; for example, data can be extracted from websites, telecom operators, governments and factories. The types of data sources are various, including pictures, video, text, and voice audio.

Velocity: Velocity implies that data generation is extremely fast. Also, due to the fast development of hardware and software, data loading and data processing are extremely speedy. Therefore, Big Data can always be considered in real time.

Veracity: Veracity implies that the quality of the obtained data is extremely different and variable. This will impact the accuracy of data analysis as well as the efficiency of the data application [2].

2.2 *Big Data infrastructure: Hadoop and Spark ecosystem*

Hadoop and Spark are the keys to the Big Data infrastructure.

Initially, Doug Cutting, who is the founder of the Apache Lucene project, designed the open-source internet text searching engine in 2002. This was the first product of Hadoop.

In 2004, the Apache Nutch Committee designed the Nutch distributed file system (NDFS). NDFS is the initial version of the Hadoop distributed file system (HDFS). In 2004 and 2005, the Nutch Committee researched and designed the open-source Google MapReduce.

Generally speaking, Hadoop is an Apache open-source framework. Based on this framework, which employs traditional low-complexity programming models, the distributed procession of large datasets, across computer clusters, becomes possible. Therefore, the benefit of Hadoop is that both programmers and small-scale companies can design their distributed programs, which can utilize cluster power to implement the storing and computing jobs, without being aware of the lower-level implementation of the distributed system [3]. To the best of the authors' knowledge, Hadoop has a number of advantages, including robustness, reliability, extensibility, computational performance, low encoding bias, and low expenditure. Recently, Hadoop has been used in the mainstream as the basis of the enterprise data mining system.

From Fig. 2, it can be seen that the key to how Hadoop works lies in the following four modules [3]:

- **YARN:** YARN is in charge of job scheduling. Besides, YARN is also utilized to manage the valuable resources of the cluster.
- **Hadoop Common:** The utility of this module contains Java libraries and utilities for other Hadoop modules, while providing the OS level abstractions, files and scripts to run Hadoop.
- **Hadoop Distributed File System (HDFS):** It is a distributed file system allowing high-throughput access to the application data.
- **MapReduce:** MapReduce is a YARN-based framework for parallel processing of large datasets. Therefore, MapReduce can be used to implement distributed computation, with promising results [4].

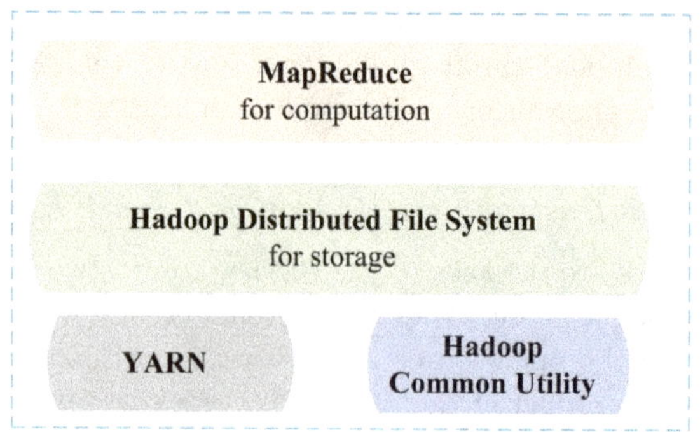

Fig. 2.　Basic modules of the Hadoop framework.

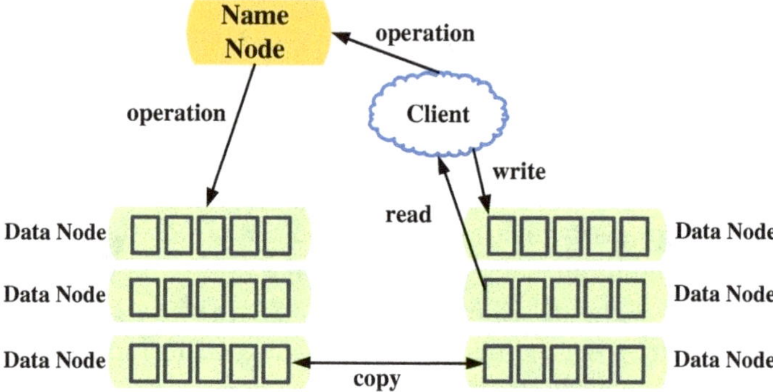

Fig. 3.　HDFS architecture.

By employing the basic modules mentioned above, the Hadoop system
has the key capabilities of distributed storage and distributed computing.
We will introduce two key modules, HDFS and MapReduce, in the following
sections.

2.2.1　*Hadoop distributed file system*

One of the variety of modules and technologies in the Hadoop ecosystem,
HDFS is the basis for the distributed data storage system. As is illustrated
in Fig. 3, HDFS employs master and slave modes [5].

Generally speaking, the NameNode software is equipped in the master
server. HDFS employs this master server to take charge of the file system
namespace to control file access.

HDFS equips DataNode software in the slave server. After receiving the request from the user/client, the slave server performs read–write operations including block creation, deletion and replication on the file systems under the instructions of the NameNode.

In HDFS, the equipped software is the key to identifying and distinguishing between the master server and the slave server. This is because the software will decide the corresponding function and work for the master server and the slave server.

In the namespace of HDFS, a specific file is split into a variety of file blocks. Then, a series of DataNodes store those file blocks in a distributed manner. In the master server, the NameNode presents the mapping of blocks to the DataNodes of slave servers. In the slave server, the DataNodes are in charge of reading the files and writing on them. Besides, they are in charge of the block-related work to be done on the file, following the master server's instructions.

2.2.2 *MapReduce*

MapReduce is a programming model designed for parallel computation on datasets with a big volume (larger than 1 TB). To reach this objective, both Map job and Reduce job (see Fig. 4) are employed to achieve the whole procedure of parallel computation [6]. More specifically, the Map job transforms the original data to other types of datasets, which consist of key-to-value pairs. Then, after receiving the processed data from the Map job, the Reduce job tries to consolidate the data to form a smaller set of tuples.

From the above description, it can be seen that the MapReduce framework provides a complex but refined parallel computation framework. This framework can automatically accomplish the parallel processing needed for computing jobs, and it can automatically identify/divide the computing

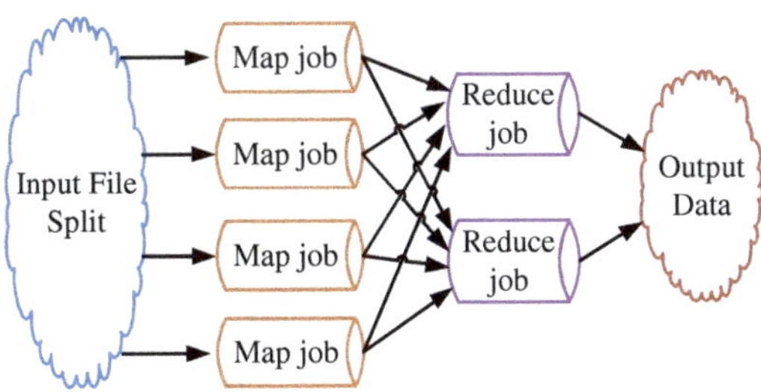

Fig. 4. Map job and Reduce job.

data and computing jobs. Furthermore, the MapReduce framework can do automatic job assignment and job execution as well as compute the collection of results on cluster nodes. Hence, the MapReduce framework can assist parallel computing to implement the complicated processes of the underlying system, including distributed data storage, data communication and fault-tolerant processing. Therefore, by employing the MapReduce framework, software designers can significantly reduce their workload. Google utilizes the MapReduce framework for more than 10,000 projects, especially large-scale computation, ML and so on.

2.2.3 *Spark framework*

Spark originates from the Big Data computation platform of the UC Berkeley AMP Lab. In 2010, Spark became open source. In the 2014 Daytona Gray Sort 100TB Benchmark competition, Spark broke the sorting algorithm record previously maintained by Hadoop MapReduce. In 2014, Spark became the top project of Apache, and then Spark 1.0.0 was released officially.

Apache Spark is a widely used Big Data processing framework which is designed for high-speed, complex analysis, for ease of use and for a large-scale dataset. Technically, Spark is designed as an in-memory parallel computing framework.

Figure 5 describes the Spark deployment diagram. In order to effectively assist the various functionalities of the Spark framework, Spark employs a series of open-source libraries and packages. For example, Spark employs MLlib for ML, SparkSQL as a distributed structured query language (SQL) query engine, and Spark Streaming for the processing and computation of streaming. Besides, Spark employs BigDL as a distributed deep learning framework as well as GraphX for distributed graph computation and graph mining [7].

In addition, many programming languages are available on Spark, including Java, R and Python. For example, SparkR is employed as a lightweight R programming language on Spark. Spark employs the org.apache.spark.api.java package for Spark Java API. As shown in Fig. 5, Spark is also compatible with many key components, for example, Flink, Hive, MapReduce, Mesos, and Storm.

2.2.4 *Spark principles*

For Hadoop MapReduce, there exists a shuffle operation between the Map job and the Reduce job. This shuffle operation is extremely time-consuming,

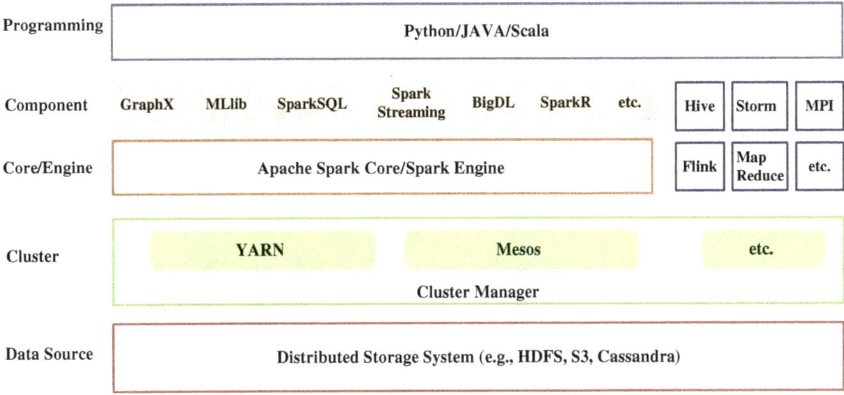

Fig. 5.   Spark deployment diagram.

since it needs duplicate data from different nodes in the HDFS. This indicates that the performance of the disk I/O is slower than the computation itself.

To effectively deal with the above time-consuming problem, Spark employs the resilient distributed dataset (RDD), where RDD is the abstraction of both data and computation in Spark. Technically, RDD employs in-memory storage technology to keep the partitioned data. This collection of partitioned data is unchangeable, but it can be operated on. RDD operations include transformation and action.

(1) **Transformation Job:** Transformation generates new RDDs from one or more RDDs.
(2) **Action Job:** Action generates the final result from an RDD.

An RDD can be generated in two ways. The first way is to read directly from the data source. For the second way, Spark uses SparkContext to generate one or more RDDs; then Spark employs a series of transformation jobs to generate the final RDD. Lastly, Spark employs the Action job to get the final computation result from these final RDDs. Compared with the Map job and the Reduce job in MapReduce, Spark can employ transformation and action jobs to simplify program development significantly.

Figure 6 illustrates the Spark application running process. In the first step, the client initializes the Spark application request and initializes the driver. In the second step, the driver program creates SparkContext. In the third step, SparkContext directly communicates with the cluster manager (e.g., Hadoop Yarn, Apache Mesos) and gets the required cluster

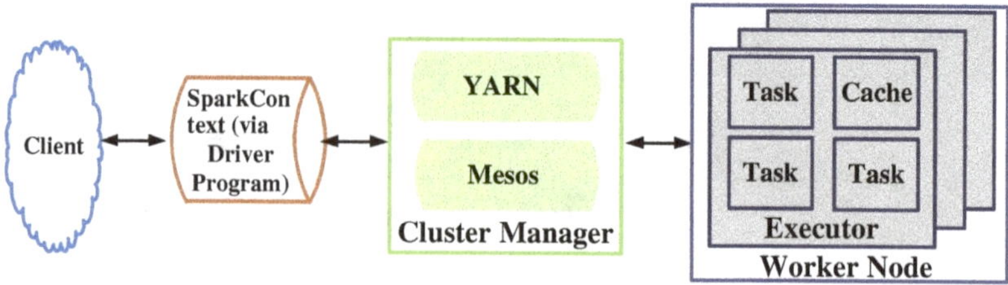

Fig. 6. Spark application running flow.

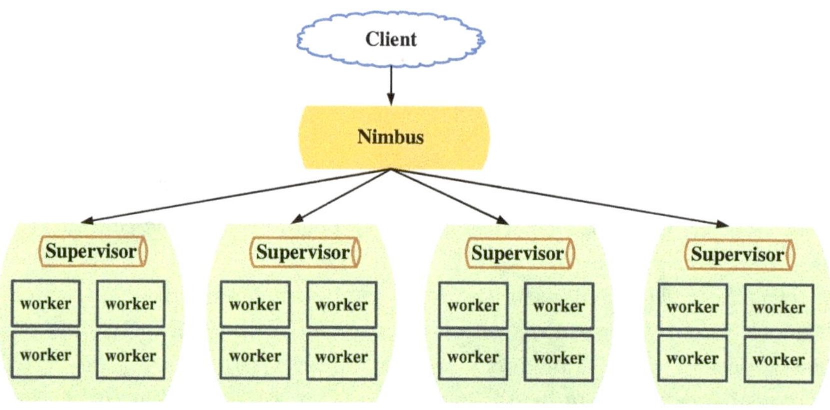

Fig. 7. Storm application running flow.

resources. In the fourth step, SparkContext acquires executors on the worker nodes, which can provide the Spark application program with distributed computation and data storage function. SparkContext sends the code of the Spark application program to each executor correspondingly, and finally, executors will execute the tasks to run the computation.

The preceding sections have introduced the widely used Big Data infrastructure, including Hadoop and Spark. Also, Storm is a promising and widely used Big Data technology. Storm is a distributed real-time Big Data processing structure. Figure 7 describes how the Storm application runs [8].

- **Nimbus:** Nimbus is in charge of Storm resource allocation and task scheduling, namely, it plays the role of a master in Storm.
- **Supervisor:** Supervisor takes the role of a slave in Storm. Supervisor receives the tasks allocated by Nimbus. As shown in Fig. 7, Supervisor

consists of a few workers, and the supervisor manages these workers in Storm.

- **Worker:** Worker takes the role of the working process in Storm. Each Worker consists of a few tasks.

2.3 *Data generation*

Figure 8 describes the vehicular Big Data value chain. The vehicular Big Data value chain includes four stages, namely vehicle data generation, data aggregation and transmission, data storage and data application. The four stages will be explained in detail in the following sections.

Data generation is the first stage of the vehicular Big Data value chain. For the vehicle industry, there exists a series of scenarios to generate vehicle data. A typical method is V2V communication, in which there is continuous signaling and information transmission among adjacent vehicles. These signaling and information data are used for vehicle control and guiding as well as communication.

For the data generation stage, the vehicular Big Data sources can be generally categorized into five types of sources, namely vehicle driving data, transport management official data, auto manufacturers and service providers' data, app data and telecom operator data, as shown in Fig. 9.

(1) **Vehicle Driving Data:** It is obtained directly from the driving data of the vehicle itself. Some types of driving data are taxi data, tachograph data, vehicle location data and vehicle navigation data.

(2) **Transport Management Official Data:** The statistics and monitoring data are collected from the transport management department. These data include monitoring camera data (via highway and road cameras), official smart transport card data and parking data.

(3) **Auto Manufacturers and Service Providers' Data:** These data include a driver's personal information, for example, their age, gender, job and home address. In addition, auto manufacturers also have

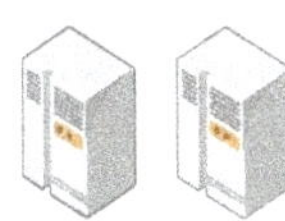

Data generation   Data aggregation & transmission   Data storage   Data application

Fig. 8.   Vehicular Big Data value chain.

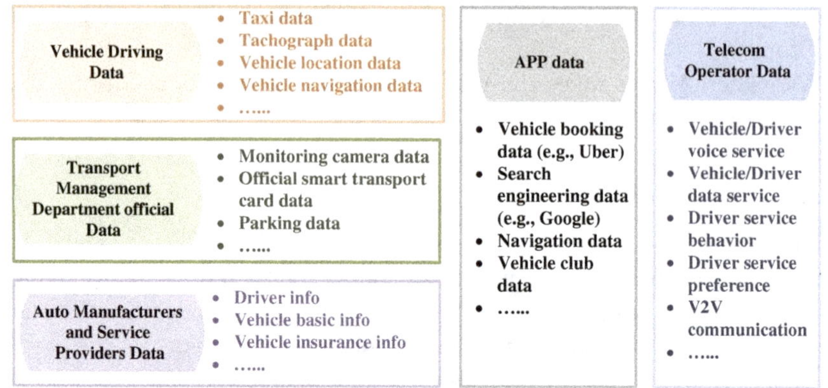

Fig. 9.   Five categories of vehicle data sources.

the driver's basic information, for example, vehicle brand, vehicle model and vehicle volume. Besides, vehicle insurance information and transportation service information are also collected by the service providers.

(4) **App Data:** There are a number of vehicle-related apps. More specifically, vehicle-booking-related apps (e.g., Uber, DiDi) collect the driver's location and regular trace data. Navigation apps can also collect the driver's location and trace data. Besides, some drivers also use search engines (e.g., Google, Baidu) and vehicle club apps; search engine and vehicle club apps also collect both driver data and vehicle data.

(5) **Telecom Operator Data:** In a V2V communication scenario, the telecom operator employs the mobile network to obtain signaling and information transmission from adjacent vehicles. In the vehicle-to-X scenario, the telecom operator can also collect data via the worldwide mobile network. Besides, telecom operators can also collect the driver's voice service and the driver's data service as well as the driver's detailed service behavior and preference.

2.4   *Data aggregation and transmission*

In Section 2.3, we introduced five types of vehicle data sources. Correspondingly, there are a number of ways to transmit these generated data. For example, vehicle-driving data can be transmitted via the V2X network. Both transport management department data and auto manufacturers'/service providers' data can be transmitted to the data center via a wired network and an optical fiber network. This is because fiber can guarantee high-speed transmission with low transmission-error ratio. In many mobile scenarios, app data and telecom operator data can be transmitted via a mobile network, for example, 2G GSM, 3G WCDMA, 4G LTE/LTE-Advanced and 5G. Also, both WiFi and Bluetooth are also popular for data transmission due to the low cost. For details of data transmission techniques, see relevant chapters that discuss V2X/mobile networks.

In order to address the data aggregation, the extract, transform and load (ETL) process is widely used. The ETL process consists of three stages, namely the extract stage, the transform stage and the load stage.

Each specific vehicle-related Big Data platform/application has a series of data source platforms with different data/file structures. Initially, the ETL technique extracts the data/file from these different data source platforms. In most scenarios, this data extraction stage is the key to ETL. This is because the completeness of important data fields directly impacts the following transform stage and load stage. Furthermore, if some important data fields/attributes are discarded in the extract stage, problems will arise during the data modeling and data mining and this might even lead to the failure of the whole project. There are many formats of data sources, for example, extensible markup language (XML), relational databases, HTTP web spidering data format, figure data format, etc. After the extract/covert/sort/split substeps of the extract stage, the data with different formats will be extracted from the different data sources and then converted into a uniform format.

After the extraction of data from different data source platforms occurs, the transform stage of ETL begins. In this transform stage, ETL will transform the extracted data into data of an appropriate format (e.g., preload format (PLF) for data warehouse) for data storage, or structure the data with the objective of data analysis and mining. Also, this transform stage plays a key role in data cleaning. A series of cleaning mechanisms and models ensures that only the appropriate extracted data is permitted to pass to the load stage (the third stage of ETL). In ETL, the transform stage comes with some challenges. For example, in the scenario of multiple platform interaction, the character sets may not be compatible and their availability cannot be guaranteed for all platforms [9].

After the transform stage, we come to the last stage of ETL, namely, the load stage. In this stage, the transformed data will be loaded into the final target. Due to the various ETL requirements, and the differences between scenarios, the final target can be a data warehouse, a target database table or a flat file. Similarly, the frequency of the load stage may vary under different scenarios and different requirements. For example, in some scenarios, the load stage for newly added extracted data implements at short intervals, namely, every minute. However, in other scenarios, the newly added extracted data may be loaded to the rational database table at relatively long intervals, namely, hourly or even daily. The time interval of the ETL process is one of the strategic design choices made on the basis of the business model and of realistic application requirements. More complicated Big Data platforms can maintain the historical audit trail changes during the loading of data in the data warehouse [9].

2.5　*Data storage*

The objective of data storage is to analyze the data characteristics of different sources. Then data storage considers the data collection method used by the service system and middleware. Correspondingly, there are a number of database types which can be used to store the vehicle data. The widely used databases can be generally divided into five categories [10–14]:

(1) Conventional relational databases (e.g., Oracle, SQL Server, MySQL, DB2);
(2) NoSQL databases (e.g., Redis, MongoDB, Couchbase);
(3) Hadoop ecosystem databases (e.g., HBase, Hive, Impala);
(4) MPP databases (e.g., Vertica, Redshift, Greenplum);
(5) Others (e.g., full-text retrieval database (SolrCloud, CNKI)).

2.5.1　*Conventional relational databases*

Relational data refers to the data which are expressed by the relational mathematical model. In the relational mathematical model of data, data are in the form of a two-dimensional table. More specifically, the relational mathematical model organizes data into one or more tables (or "relations") of columns and rows where every row can be identified by a unique key.

As for the relational database, it is a digital database created on the basis of the relational mathematical model of data (e.g., a collection of views, stored procedures, indexes). Generally, a relational database contains two-dimensional tables. The tables contained in relational databases are

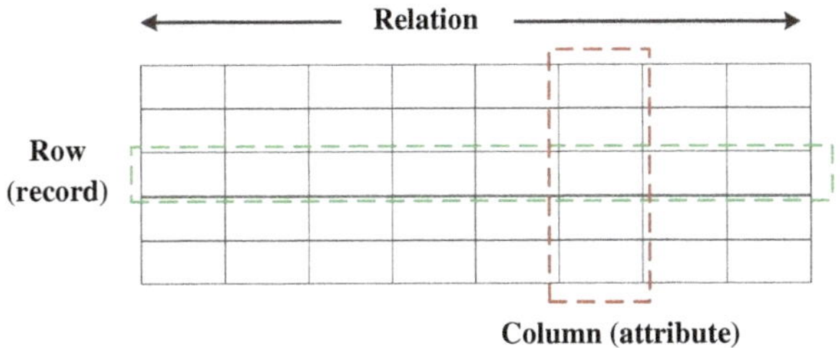

Fig. 10.   The basic idea of the relational mathematical model of data.

correlated, and the correlations are mainly realized by the reference relationship embodied by the primary key and the foreign key.

Figure 10 describes the basic idea of the relational mathematical model of data. In the relational mathematical model of data, each row represents a specific record (e.g., a specific driver's basic information). Each column is called as a specific attribute (e.g., the gender of the driver). Based on the above relational mathematical model of data, in the relational database, the rows indicate entries of that type of entity, and the columns denote values attributed to that entry. In this way, we can employ a relational database to store the rational data. Furthermore, we can employ the SQL for querying and managing the rational database.

2.5.2 *NoSQL databases*

It is generally known that some data source platforms may generate semi-structured and non-structured data. It is difficult to store and manage this data with the conventional rational databases. A NoSQL database can effectively maintain these semi-structured and non-structured data, and even large-scale structured data as well.

There are many strong points of NoSQL databases, for example, high extendibility, superior performance, flexible data model and open source. Also, the document-oriented category, the column-family category, the graph category and the key-value category are the four typical categories of NoSQL databases. There are many products of NoSQL databases, among which Redis and MongoDB are widely used in both academia and industry.

2.5.3 *Hadoop ecosystem databases*

As mentioned in Section 2.2.1, Hadoop employs HDFS to store files, which are written to HDFS by a single writer in an append-only pattern.

Therefore, many SQL-on-Hadoop systems fall short of being capable of transactions. For example, Impala falls short of being capable of transactions. Hive has only a little capability when it comes to performing transaction, and does so at the expense of reduced query stability and performance [3]. There are many popular products of the Hadoop ecosystem database, including SparkSQL, Impala and Hive.

2.5.4 *MPP databases*

MPP stands for massive parallel processing. In the non-shared cluster of a database, each node has a specific/independent disk storage system and a memory system. Also, service data are allocated to different nodes from the database model and application characteristics. Each data node connects via a dedicated network or a business network [15, 16]. In this way, all data nodes collaborate and provide database services as a whole.

From the above analysis, in the MPP database, data are normally stored locally in each node. The data are managed by a distributed file system. Technically, MPP databases are a low-cost and effective way to provide Big Data infrastructure. Many MPP database products are widely used in the industry, including Nettezza (IBM), Asterdata (Teradata), Greenplum (EMC), Vertica (HP) and Redshift (Amazon) [17, 18].

2.6 *Data application*

In recent years, vehicular Big Data have been widely researched and applied in many countries.

In Singapore, the land area is limited, with a large population. The government collects vehicular Big Data, especially the smart card data and customers trace information, to optimize the public transport resources, including shuttle bus, express and taxi. Furthermore, the government employs the data mining-based public transport multi-dimensional operation to improve the efficiency of public transport.

In some cities of the USA, vehicles and companies, as well as customers, are integrated via wireless vehicle networks. Then, the transportation department employs the collected data of vehicle networks to implement intelligent transport management and vehicular smart control.

In China, vehicular Big Data is utilized to assist the government in constructing the transportation infrastructure and optimizing the transportation resources, e.g., smart parking, highway planning.

In this section, we introduce a Big Data system for precise vehicle business operation and advertisement. As described in Fig. 11, the system

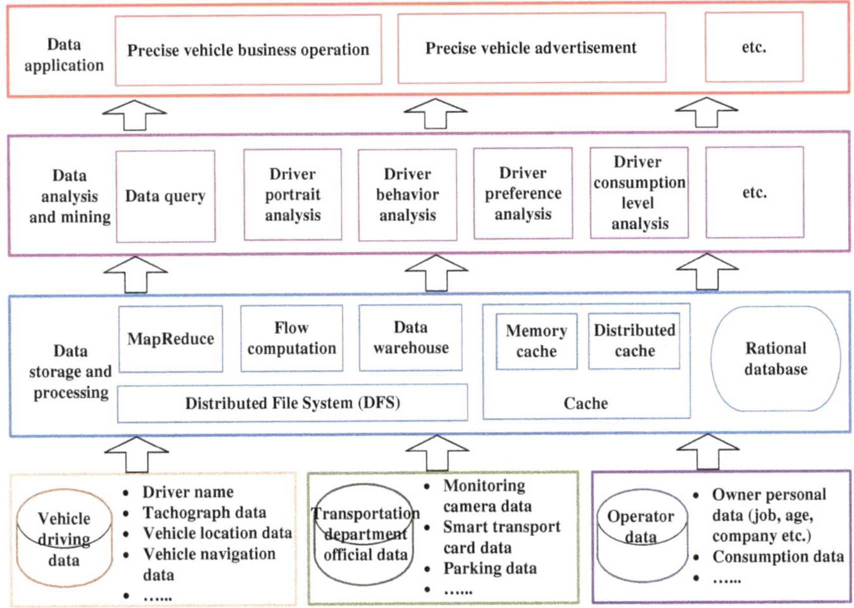

Fig. 11. Big Data system for precise vehicle business operation and advertisement.

includes four stages, including data extraction, data storage and processing, data analysis and mining, and data application.

In the data extraction stage, this system extracts vehicle driving data (e.g., driver name, tachograph data, vehicle location data, vehicle navigation data) and transportation management official data (e.g., monitoring camera data, smart transport card data and parking data). In addition, operation data are also collected in this stage, including the owner's personal data (job, age, company) and consumption data.

The data storage and processing stage are more related to the system's infrastructure. Both the distributed file system and the rational database store the collected data/files. Then, these data/files are processed via the computation engine, including MapReduce, flow computation and data warehouse.

In the data analysis and mining stage, this system queries useful data and draws the 360-degree portrait of each driver. Besides, this system also analyzes each driver's behavior and preference as well as consumption level. These analyses and mining results will be output to support the data-application stage.

In the data application stage, this system employs a series of ML algorithms for precise vehicle business operation [14]. Based on the driver's preference and consumption level, this platform mines potential buyers of automotive products and then pushes suitable advertisements (e.g., Google, Facebook and WeChat) to these potential customers.

3 Background of AI and ML

3.1 *What are AI and ML?*

In computer science, AI research focuses on creating intelligent machines (or agents) which work and react like humans, i.e., they perceive the environment and take actions to maximize their chances to achieve their goals [19] successfully. After more than half a century of development, AI has become an essential part of the IT industry and has been widely used in many fields. The significant problems faced by AI researchers are learning, reasoning, natural language processing, perception, knowledge representation, general intelligence and so on.

ML [20] is a fundamental part of AI. Unlike traditional AI research which mainly focuses on introducing hand-designed knowledge to intelligent agents, ML aims at designing computer algorithms which automatically learn and improve themselves through experience. Mitchell [21] proposed a commonly accepted, more formal definition of ML algorithms as follows.

Definition 1.1 (Machine Learning). A computer program is said to learn from experience E concerning some class of tasks T and performance measure P if its performance at tasks in T, as measured by P, improves with experience E.

Typically, the experience can be represented by data examples. Depending on whether the data examples have associated outputs (labels), we may divide ML tasks into the following four groups:

- **Supervised Learning:** Each example input is associated with the output, and an ML algorithm aims at learning a general rule which can be used to map each example input to its corresponding output.
- **Unsupervised Learning:** There are no outputs associated with the example inputs; the learning algorithm has to uncover the hidden structure of the inputs by itself.
- **Semi-supervised Learning:** Only some of the example inputs have corresponding outputs; the algorithm has to take advantage of both labeled and unlabeled examples.

- **Reinforcement Learning:** The output associated with each input is provided only through feedback (rewards or punishments) to the agent's actions in a dynamic environment, and the learning algorithm aims to maximize the long-term rewards.

ML tasks can also be divided based on the output of a learning algorithm, resulting in the following five classes:

- **Classification:** Divide the inputs into two or more classes.
- **Regression:** Based on the inputs, predict their outputs which are continuous values rather than class labels.
- **Clustering:** Divide the inputs into several groups without the need for information on their outputs.
- **Density Estimation:** Estimate input example distribution in some probability space.
- **Dimensionality Reduction:** Map the feature space of input examples into a lower dimensional feature space, such that they can be more efficiently computed.

3.2 *Supervised learning*

In supervised learning, example inputs are associated with the corresponding outputs, and the learning algorithm aims at learning a function from the examples so that the learned function can map each example input to its corresponding output [22]. For each training example, i.e., input–output pair, the input is typically modeled as a feature vector while the output (discrete or continuous) is a desired value (label) corresponding to the input. For classification tasks, the outputs are typically discrete values corresponding to different class labels, while for regression tasks, the outputs are continuous values. After a learning model is trained, it can be applied to new examples to predict their outputs. Formally, supervised learning can be defined as follows.

Definition 1.2 (Supervised Learning). Given a set of N training examples $\{(x_1, y_1), \ldots, (x_N, y_N)\}$, where x_i is the feature vector of the ith example input and y_i is its corresponding output (label), a supervised learning algorithm aims to learn a function $g : X \to Y$, where X is the input space and Y is the output space. g is evaluated by a scoring function $f : X \times Y \to \mathrm{R}$, such that g is defined as returning the y value with the highest score: $g(x) = \arg_y \max f(x, y)$.

Two factors should be considered in designing f and g, i.e., empirical risk minimization and structural risk minimization. The first seeks to fit the

Table 1. Comparison of classical supervised learning algorithms.

Algorithm	Task type	Interpretable results	Training speed	Prediction speed	Parametric
Linear regression	Regression	Yes	Fast	Fast	Yes
Logistic regression	Classification	Somewhat	Fast	Fast	Yes
SVM	Classification	Somewhat	Moderate	Fast	No
KNN	Either	Yes	Fast	Depends	No
Naive Bayes	Either	Somewhat	Fast	Fast	Yes
Decision trees	Either	Somewhat	Fast	Fast	No
Random forests	Either	A little	Slow	Moderate	No
AdaBoost	Either	A little	Slow	Fast	No
Neural networks	Either	No	Slow	Fast	Yes

Note: See https://www.dataschool.io/comparing-supervised-learning-algorithms/ for a more comprehensive comparison.

training data as well as possible while the latter tries to avoid overfitting by introducing a penalty function.

In the past few decades, a large number of studies have been proposed to investigate supervised learning algorithms. Table 1 offers a comparison of the classical approaches [23].

3.3 *Unsupervised learning*

Unlike supervised learning where each example input is associated with an output, i.e., examples are labeled, unsupervised learning can only rely on "unlabeled" examples, and the learning algorithm has to uncover the hidden structure of the inputs by itself. One difference between unsupervised learning and supervised learning is that, for unsupervised learning, there is no straightforward way to evaluate the accuracy of the algorithm due to the lack of labels. Typical unsupervised learning approaches include the following categories [24]:

- **Clustering:** Clustering, also called cluster analysis, aims to group or segment a set of objects into subgroups or "clusters," so that those within each subgroup (namely, cluster) are more tightly related to each other while those assigned to different subgroups (clusters) are sparsely related. Representative clustering algorithms are K-means clustering, hierarchical clustering, Gaussian mixture models (GMM) and so on.
- **Dimensionality Reduction:** It is defined as the procedure of reducing the number of random variables by computing a set of primary variables.

In ML, it usually serves as a preprocessing step for faster and simpler model training. Algorithms of dimensionality reduction include singular value decomposition (SVD), principal component analysis (PCA), non-negative matrix factorization (NMF), etc.

- **Neural Networks:** The two major neural network models for unsupervised learning are autoencoders and the self-organization map, where the former learns to compress data from the input layer into a shortcode while the latter produces a low-dimensional (typically two-dimensional), discretized representation of the input space.

3.4 *Semi-supervised learning*

Semi-supervised learning [25, 26] falls between supervised learning (with all examples labeled) and unsupervised learning (without any labeled examples). Because in some real-world applications, labeling a significant amount of data is often difficult, expensive, tedious or time-consuming, it demands special devices or needs the efforts of human experts. On the contrary, unlabeled data can be more easily and inexpensively obtained. So, learning algorithms should be able to utilize both labeled and unlabeled examples of building better learners, rather than using each type alone.

Semi-supervised learning can be further classified into transductive learning and inductive learning. The former seeks to infer the correct labels for the unlabeled training data while the latter aims at predicting labels for future examples.

Typical algorithms of semi-supervised learning include low-density separation, generative models, heuristic approaches, graph-based methods and so on.

3.5 *Reinforcement learning*

RL [27], as an essential research field of ML, focuses on how intelligent agents interact with a dynamic environment to maximize the cumulative reward, as shown in Fig. 12. Base reinforcement can be formulated as the following Markov decision process [28]:

- a set of agent and environment states, S;
- a set of actions, A, of the agent;
- state transition probability, $P_a(s, s') = \Pr(s_{t+1} = s' | s_t = s, a_t = a)$, i.e., the possibility of transition from the state s to the state s' under action a;

     L. Zhao, L. Xu & J. Shang

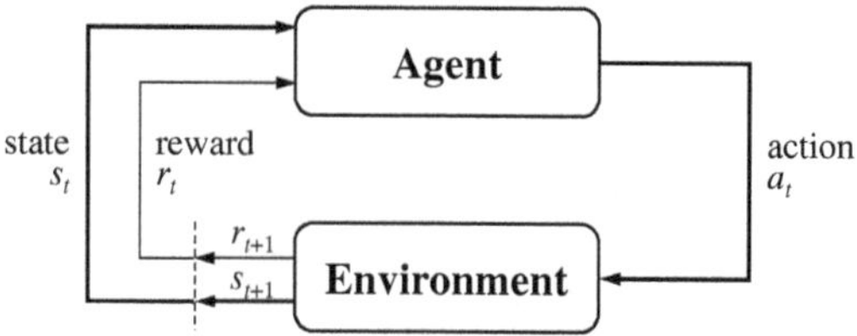

Fig. 12. The agent–environment interaction in RL [27].

Table 2. Comparison of RL algorithms [28].

Algorithm	Description	Action space	State space	Operator
Monte Carlo	Every visit to Monte Carlo	Discrete	Discrete	Sample means
Q-learning	State-Action-Reward-State	Discrete	Discrete	Q-value
SARSA	State-Action-Reward-State-Action	Discrete	Discrete	Q-value
DQN	Deep Q Network	Continuous	Continuous	Q-value
DDPG	Deep Deterministic Policy Gradient	Continuous	Continuous	Q-value
A3C	Asynchronous Actor-Critic Algorithm	Continuous	Continuous	Q-value
NAF	Q-Learning with Normalized Advantage Functions	Continuous	Continuous	Advantage
TRPO	Trust Region Policy Optimisation	Continuous	Continuous	Advantage
PPO	Proximal Policy Optimisation	Continuous	Continuous	Advantage

- $R_a(s, s')$ the immediate reward after the transition from s to s' under a;
- rules that describe what the agent observes.

As shown in Fig. 12, in RL, the agent interacts with its dynamic environment. At each time t, the agent takes action a_t based on the current state s_t and reward r_t, then the environment moves to the next state s_{t+1}, and the new reward r_{t+1} is determined. The learning purpose of the agent is to accumulate as many rewards as possible; to achieve this, the agent must consider long-term and short-term rewards simultaneously. Table 2 summarizes several classic RL algorithms [28].

4 AI in Communications for Smart Mobility

Communications and networking are the key technologies to enable the connectivity of vehicles, people and even freights on the roads of the smart city [29, 30]. To ensure efficient communications for smart mobility, AI and its subset ML should be involved in communications. In order that we may fully consider the applications of AI in communication, it is necessary to discuss AI in all five layers of communication, from bottom to top, namely, the physical layer, the link layer, the network layer, the transport layer and the application layer (see Fig. 13). Therefore, in this section, we would like to present the various AI technologies in five layers.

4.1 *AI in the physical layer*

Cognitive radio (CR) is a typical application of AI technology in the physical layer of mobile networks. CR is a challenging technology that improves the utilization of the channel spectrum [31]. In particular, CR is used for mobile systems such as cellular networks to utilize the unused spectrum, and its commercialized application is the third-generation mobile communication technology (3G).

With the development of software (SW) and hardware (HW) for communication devices, the applications (e.g., high-definition video, online gaming, augmented reality) could consume more bandwidths. To meet the

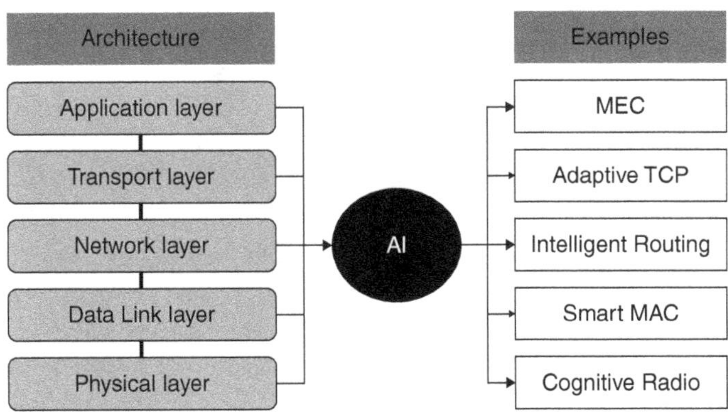

Fig. 13. Applications of AI in the five-layer network.

increasing demand for bandwidth, CR has been found to be a promising solution as it allocates the spectrum access dynamically. Smart mobility is one of the most critical scenarios in the upcoming fifth-generation mobile communication technology (5G). In particular, the internet of things (IoT) and its variant, the vehicle of things (VoT), are both strongly supported by the future 5G standards, in which the entities of IoT and VoT are also part of smart mobility as are vehicles, people and other mobile devices. The applications of CR are cognitive networking [32], self-organizing networks [33], cooperative relaying and networks [34], vehicular networks [35] and so on.

CR is considered as a technology for the intelligent wireless communication system in which it can monitor the spectrum usage and the possible use of the unused spectrum and assign it to unlicensed users (secondary users, SUs) without bothering about the usage of the licensed users (primary users, PUs) [36]. We can commonly divide the spectrum resources of CR into two categories: unlicensed bands and idle licensed bands (i.e., spectrum holes). Most available spectrum resources are below 6 GHz. Therefore, to achieve a higher data rate, the 5G system extends the band capacity by considering the possible spectrum from 1 GHz to 100 GHz. However, this range is still minimal. It is worth mentioning that most licensed spectrum resources are not fully utilized. Therefore, when radio resources are scarce, we can apply the CR technology to reuse the spectrum with minimum influence on the current spectrum allocation policy. This method could improve the utilization rate and further increase the data rate. Spectrum sharing includes four steps: spectrum sensing [37], allocation, handoff and access. First, in the spectrum-sensing phase, CR methods are used to detect whether a PU or spectrum hole is available. Then, according to the designed spectrum allocation rules, the available spectrum is allocated efficiently with minimum conflicts. By applying the access algorithm, after the PUs have been given access to the allocated bands, the SUs can access the spectrum efficiently. The SUs should switch to other spectrums whenever the PU intends to obtain an SU-engaged hole, or the position of the SU changes or the frequency band of the SU cannot meet the requirements.

To efficiently accomplish the four steps above, AI is widely used as a cognitive engine (CE) [38] to analyze and classify the different situations and make proper decisions. The artificial neural network is the first kind of AI technology used in the CR we would like to introduce in this chapter. The ANN has been used for setting radio parameters adaptively [39–41]. The radio parameters can be determined by applying the ANN to meet the goals of maximum throughput, minimum transmission power and BER [39]. The ANN-based distributed optimization algorithm is proposed for a cognitive wireless cloud [40]. Besides ANN-based CR, metaheuristic algorithms [42] can be used to obtain solutions to computationally hard problems. For instance, the genetic algorithms (GAs) have been adopted to solve optimization problems by adjusting and configuring the CR dynamically, in particular, in the changing wireless mobile environment [43, 44]. For example, GA is applied to adapt the radio parameters of the software-defined radio [45].

Further, as a statistical model for analyzing and explaining random events or patterns, the hidden Markov model (HMM) can be used to observe the process of the CE or prediction. For example, the HMM is applied to model an online model for the wireless channel of CR [46]. Besides, HMMs are also used for spectrum sensing in CR [47]. The HMM is proposed to train by using GA where the training data are from the broadband channel sounder.

4.2 *AI in the link layer*

AI has been widely applied to enhance the performance of the MAC layer, including the applications of RL, fuzzy logic, neural networks and so on.

First, RL is the most promising method used to refine the MAC layer protocol. In [48], with the observation, action and reward mechanisms, Q-learning, as an RL, is adopted in Aloha-QIR to minimize retransmissions and collisions while improving throughput and efficiency. Also, in order to improve throughput and latency, RL-MAC [49] is also proposed with five main components, namely reward formulation, early sleeping avoidance, Q-learning, overhearing and collision avoidance. Among the above five components, reward formulation and Q-learning functions are the most RL-related components and involve mechanisms, such as dividing the rewards, and updating and selecting the next action. The security level is expected to improve in WSNs if SVM is used, where [50] intends for attacks to be detected at the medium control of the MAC layer.

Second, fuzzy logic is applied to improve CSMA/CA by reducing the packet loss ratio [51]. The fuzzy logic control can adjust the parameters of the contention window and backoff exponent by considering throughput and workload.

Third, the ANN is applied to detect the distributed denial of service (DDoS) attacks and turn down the MAC layer whenever the output of the model is over a threshold [52]. In [53], Qiao *et al.* apply a neural network

(NN) to allow selection and adaptive switching between the different MAC protocols, DCF and TDMA, according to the network load.

4.3 *AI in the network layer*

The applications of AI in the network layer are mainly in the areas of refining the routing protocols, in particular for vehicular networking [54] and WSNs. Typically, two types of AI algorithms are adapted here. One is the swarm intelligence (SI) optimization while the other is ML [55, 56]. For SI, the whale optimization algorithm is applied to search for the best route for data packets [54].

For ML, the RL technique, Q-learning, is applied to search the optimal routes for underwater WSNs in order to minimize the energy consumption which may be caused by conventional shortest path algorithms like Bellman–Ford [55]. Another RL-based routing approach [56] has also been proposed to create efficient communication between roadside sensor nodes and vehicles. Reference [56] also applies Q-learning, while considering the reward of energy and hop counts. From this point, we can observe all types of ML methods, and RL and its representative Q-learning are most frequently used to solve the routing problem currently. Besides using ML to select the optimal packet delivery route, ML methods such as extreme learning machines [57] could also be used to predict the position of the neighboring nodes [58]. Besides RL, the Bayesian classifier has also been applied to predict the behavior of nodes to help the routing protocol during the selection of intermediate nodes [59]. ML is also used to enhance the quality of forwarder selection validators and routing metrics [60–62].

For SI, many swarm intelligent optimization algorithms have been used to improve the routing algorithms since the swarming group has the characteristic of seeking optimal paths. AntNet [63] works by applying the ant colony optimization (ACO) algorithm which sends ants to sense the possible routes from source to destination and selects the next-hop forwarder from the pheromone tables (routing tables). However, we can imagine that sending out ants (discovery packets) to randomly explore the network could increase the overheads. Reference [54] introduces the multi-objective auto-regressive whale optimization (ARWO) algorithm which chooses the optimal path among the multiple paths by considering the various objectives, including latency, link lifetime and distance. ARWO predicts the traffic density of the traffic and the average velocity of the vehicles to enhance the routing performance. Although the routing overhead is lower than the reactive protocols such as AODV, it is still high, in particular when we compare SI-enabled routing protocols to location-based protocols such as GPSR. From this point, we can see that SI-enabled routing protocols are not entirely suitable for vehicular communications. However, SI such as the grey wolf optimization algorithm could be deployed to create the optimal number of clusters according to the social behavior of grey wolves in VANETs [28].

4.4 *AI in the transport layer*

In the switching network, ML has been widely used to improve the performance of TCP, e.g., by predicting the TCP throughput based on file transfer history and path properties [64, 65]. On the contrary, ML has also been adopted in many areas in the transport layer including TCP and intrusion detection. ML is used to select the initial window size and set up the value of the congestion window to optimize the network performance under TCP transmissions by applying ML [66, 67]. Reference [68] also demonstrates the RAN-assisted TCP window optimization which involves the backpropagation (BP) neural network. In [68], BP is utilized to predict the most appropriate size of TCP windows based on the context of wireless transmission. More information about the AI in TCP is provided in [69], in which Lin proposes a network resource management algorithm by applying ML methods. A model is learned as the relationship between the resource allocation and application requirements. With the model, the multipath TCP is then allowed in the heterogeneous wireless networks. The research team of Leung [70] proposed an adaptive network coding strategy based on Q-learning and logistic regression for controlling the number of packets according to the network changes.

4.5 *AI in the application layer*

Compared to other layers, the study of AI in the application layer has attracted more attention [71–73], in particular in the area of autonomous vehicles. In [73], reputation systems have been built by applying SVM to identify malicious nodes to improve the reliability of p2p applications in wireless networks.

Other technologies such as mobile edge computing (MEC) can also be used to enhance the performance of deep learning for mobile communication. In [60], Li *et al.* proposed an offloading strategy to improve performance for deep learning applications in IoT by adapting MEC servers.

5 Discussion and Conclusion

5.1 *Basic design workflow of applying Big Data and AI in communications for smart mobility*

We introduce the basic workflow of applying ML and Big Data in communications for smart mobility, including problem formulation, Big Data collection, data preprocessing and feature extraction, model construction (training and tuning), model validation and deployment [74]. In this section, we would like to explain each step in the workflow:

- **Problem Formulation:** It is necessary to make sure and formulate the problem correctly before doing anything else. The problem could be classification, regression (decision-making) or clustering. By abstracting the type of target problem, we can decide the type and volume of data that we need to collect as well as the right learning model to apply for training the data.
- **Big Data Collection:** To learn about a proper model from the massive network data, a significant amount of network data should be collected. The Big Data may include network data such as network traffic traces and logs, or non-network data such as weather, road and vehicle data [60]. For example, in [60], to construct the network traces of vehicle networking traces, the floating car data (FCD) of Beijing has been collected. Then, the network simulation has also been taken from the historical FCD data to build the labeled dataset.
- **Data Preprocessing and Feature Extraction:** In this step, we have to find the most impacted and valuable factors that have the most effect on the target network problem. By observing the proper features, this could help reduce the number of features and the processing time and size of the dataset.
- **Model Construction (Training and Tuning):** A proper ML method should be chosen regarding the size of the dataset, the type of problem and so on. By using the right ML method, a model could be learned based on the historical data. For most ML methods, we still need to run the parameter-tuning process to find suitable parameters for the most appropriate model either by using the optimization algorithm or by relying on experience.

- **Model Validation:** The test dataset should be used to evaluate the accuracy of the model. From the validation results, we can derive a guidance pattern that can show us how we can optimize the model either by increasing the volume of the training dataset or reducing the complexity of the model.
- **Deployment:** As the final step, the learning model should be implemented in the network environment by considering the practical issues. The computation complexity and energy consumptions should concern us since most mobile devices including vehicles suffer from limited computation and energy resources. For example, when adapting the model in a routing protocol as a routing metric [60], sufficient response time should be guaranteed to ensure the operation of data packet forwarding.

5.2 *Crowdsourcing for Big Data and artificial intelligence*

In the field of Big Data, crowdsourcing [75] is used as a sourcing model to collect data by involving some less specific public internet users. For the communication involved in smart mobility, crowdsourcing is even more critical since we have to collect data from vehicles, pedestrians and even infrastructures to enable effective communication. Also, most vehicular data are generated by vehicle sensors, which are costly, and it would be hard to ensure that all cars are equipped with a full set of sensors for autonomous driving in future. Therefore, there may be vehicles with different sensors on the road in future, and they shall use the data of the sensors both for autonomous driving as well as for communication. In particular, when we apply software-defined vehicular networks (SDVNs) to replace the current open system interconnection (OSI) architecture and TCP/IP architecture, the controller of the SDVNs would collect the vehicular information and operate the centralized networking. In such cases, how to collect crowdsourcing data from the participants becomes a significant topic in communication for smart mobility.

5.3 *Conclusion*

We are now in an era that is changing; the conventional living style of the present is changing to become a future involving smart cities. The application of intelligence in mobility allows human beings and goods to move efficiently from one place to another. With the adaption of communication, all entities would be connected during these movements, which would further increase the efficiency and reliability of the smart mobility scenario. For this reason, this chapter introduces the application of Big Data and AI to the communications involved in smart mobility. We show the main techniques, namely, Big Data, AI and ML, and review the existing works on how to apply Big Data and AI in mobile communications. We also briefly discuss the workflow of adopting Big Data and AI in mobile communications. The topics discussed in this chapter have attracted much attention in the last few years, and we believe the ongoing research on AI in mobile communications will still continue, covering network control, software-defined radio, routing and even communication security.

References

[1] De Mauro, A., Greco, M., & Grimaldi, M. (2016). A formal definition of Big Data based on its essential features. *Library Review*, 65(3), 122–135.

[2] Big Data. Wikipedia. https://en.wikipedia.org/wiki/Big_data.

[3] Su, F., Wang, Z., Yang, S., Li, K., Lu, X., Wu, Y., & Peng, Y. (2017). A survey on Big Data analytics technologies. In *International Conference on 5G for Future Wireless Networks*, Beijing, China, pp. 359–370.

[4] Apache Hadoop. http://hadoop.apache.org/.

[5] HDFS Architecture Guide. Apache Hadoop. https://hadoop.apache.org/docs/r1.2.1/hdfs_design.html.

[6] Arun M., & Vinod K. V. (2014). *Apache Hadoop YARN — Moving Beyond MapReduce and Batch Processing with Apache Hadoop 2*. Addison-Wesley Professional.

[7] Zaharia, M., Chowdhury, M., Franklin, M. J., Shenker, S., & Stoica, I. (2010). Spark: Cluster computing with working sets. *HotCloud*, 10(10–10), 95.

[8] Goetz, P. T., & O'Neill, B. (2014). *Storm Blueprints: Patterns for Distributed Real-time Computation*. Packt Publishing Ltd.

[9] Extract, transform, load. Wikipedia. https://en.wikipedia.org/wiki/Extract,_transform,_load.

[10] Apache Impala. Apache. http://impala.io/.

[11] Apache Spark. Apache. http://spark.apache.org/.

[12] Apache Strom. Apache. http://storm.apache.org/.

[13] Apache Hbase. Apache. https://hbase.apache.org/.

[14] Zhang, H., Zhang, L., Cheng, X., & Chen, W. (2016). A novel precision marketing model based on telecom big data analysis for luxury cars. In *16th IEEE International Symposium on Communications and Information Technologies (ISCIT)*, Qingdao, China, pp. 307–311.

[15] Karau, H., Konwinski, A., Wendell, P., & Zaharia, M. (2015). *Learning Spark: Lightning-fast Big Data Analysis*. O'Reilly Media, Inc., pp. 182–211.

[16] Babu, S., & Herodotou, H. (2012). Massively parallel databases and MapReduce systems. *Foundations and Trends in Databases*, 5(1), 1–104.

[17] Cheng, B., Guan, X., & Wu, H. (2015). A hypergraph based task scheduling strategy for massive parallel spatial data processing on master-slave platforms. In *23rd International Conference on Geoinformatics*, Qingdao, China.

[18] Lu, X., Su, F., Liu, H., Chen, W., and Cheng, X. (2016). A unified OLAPOLTP Big Data processing framework in telecom industry. In *IEEE International Symposium on Communications and Information Technologies (ISCIT)*, Qingdao, China.

[19] Legg, S., & Hutter, M. (2007). A collection of definitions of intelligence. *Computer Science*, 17–24.

[20] Jordan, M. I., & Mitchell, T. M. (2015). Machine learning: Trends, perspectives, and prospects. *Science*, 349(6245), 255–260.

[21] Mitchell, T. M. (1997). *Machine Learning*, Vol. 45(37). Burr Ridge, IL: McGraw Hill, pp. 870–877.

[22] Russell, S. J., & Norvig, P. (2010). *Artificial Intelligence: A Modern Approach*, 3rd edn. Prentice Hall.

[23] Caruana, R., & Niculescu-Mizil, A. (2006). An empirical comparison of supervised learning algorithms. In *23rd International Conference on Machine Learning*, Pittsburgh, Pennsylvania, USA, pp. 161–168.

[24] Hastie, T., Tibshirani, R., & Friedman, J. (2009). Unsupervised learning. In *The Elements of Statistical Learning*. Springer, NY, pp. 485–585.

[25] Zhu, X. (2011). Semi-supervised learning. In *Encyclopedia of Machine Learning*. Springer, Boston, MA, pp. 892–897.

[26] Hady, M. F. A., & Schwenker, F. (2013). Semi-supervised learning. In *Handbook on Neural Information Processing*. Springer, Berlin, Heidelberg, pp. 215–239.

[27] Sutton, R. S., & Barto, A. G. (1998). *Reinforcement Learning: An Introduction*. MIT Press.

[28] Reinforcement Learning. Wikipedia. https://en.wikipedia.org/wiki/Reinforcement_learning.

[29] Han, G., Jiang, J., Zhang, C., Duong, T. Q., Guizani, M., & Karagiannidis, G. (2016). A survey on mobile anchors assisted localization in wireless sensor networks. *IEEE Communications Surveys & Tutorials*, 8(3), 2220–2243.

[30] Simsek, M., Aijaz, A., Dohler, M., Sachs, J., & Fettweis, G. (2016). 5G-enabled tactile internet. *IEEE Journal on Selected Areas in Communications*, 34(3), 460–473.

[31] Mitola, J. (2000). Cognitive radio — An integrated agent architecture for software defined radio. Ph.D. Thesis, Royal Institute of Technology (KTH), Stockholm, Sweden.

[32] Thomas, R. W., Friend, D. H., DaSilva, L. A., & MacKenzie, A. B. (2007). Cognitive networks. In *Cognitive Radio, Software Defined Radio, and Adaptive Wireless Systems*. Springer, Dordrecht, pp. 17–41.

[33] Niyato, D., & Hossain, E. (2008). Spectrum trading in cognitive radio networks: A market-equilibrium-based approach. *IEEE Wireless Communications*, 15(6), 71–80.

[34] Lu, C., Fitzek, F. H., & Eggers, P. C. (2007). A cooperative scheme enabling spatial reuse in wireless networks. In *Cognitive Wireless Networks*. Springer, Dordrecht, pp. 457–471.

[35] Neel, J., & Amanna, A. (2009). Introduction to cognitive radio technology for transportation systems. In *SDR Forum Workshop Smart Communication, Transport System*.

[36] Hu, F., Chen, B., & Zhu, K. (2018). Full spectrum sharing in cognitive radio networks toward 5G: A survey. *IEEE Access*, 6, 15754–15776.

[37] Chung, W., Park, S., Lim, S., & Hong, D. (2014). Spectrum sensing optimization for energy-harvesting cognitive radio systems. *IEEE Transactions on Wireless Communications*, 13(5), 2601–2613.

[38] He, A., Bae, K. K., Newman, T. R., Gaeddert, J., Kim, K., Menon, R., & Tranter, W. H. (2010). A survey of artificial intelligence for cognitive radios. *IEEE Transactions on Vehicular Technology*, 59(4), 1578–1592.

[39] Reed, J. H., *et al.* (2005). Development of a cognitive engine and analysis of WRAN cognitive radio algorithms — Phase I. Wireless @ Virginia Tech, Virginia Polytechnic Institute State University, Blacksburg, VA.

[40] Hasegawa, M., Tran, H. N., Miyamoto, G., Murata, Y., & Kato, S. (2007). Distributed optimization based on neurodynamics for cognitive wireless clouds. In *IEEE 18th International Symposium Personal, Indoor and Mobile Radio Communications*, Athens, Greece.

[41] Zhang, Z., & Xie, X. (2007). Intelligent cognitive radio: Research on learning and evaluation of CR based on neural network. In *2007 IEEE Information and Communications Technology Conference*, Caira, Egypt, pp. 33–37.

[42] Blum, C., & Roli, A. (2003). Metaheuristics in combinatorial optimization: Overview and conceptual comparison. *ACM Computing Surveys*, 35(3), 268–308.

[43] Newman, T. R., Barker, B. A., Wyglinski, A. M., Agah, A., Evans, J. B., & Minden, G. J. (2007). Cognitive engine implementation for wireless multicarrier transceivers. *Wireless Communications and Mobile Computing*, 7(9), 1129–1142.

[44] Park, S. K., Shin, Y., & Lee, W. C. (2007). Goal-Pareto based NSGA for optimal reconfiguration of cognitive radio systems. In *IEEE Cognitive Radio Oriented Wireless Networks and Communications Conference*, Orlando, FL, USA, pp. 147–153.

[45] Rondeau, T. W., Le, B., Rieser, C. J., & Bostian, C. W. (2004). Cognitive radios with genetic algorithms: Intelligent control of software defined radios. In *SDR Forum Technical Conference*, Vol. 100, Phoenix, pp. 3–8.

[46] Rondeau, T. W., Rieser, C. J., Gallagher, T. M., & Bostian, C. W. (2004). Online modeling of wireless channels with hidden markov models and channel impulse responses for cognitive radios. In *IEEE MTT-S International Microwave Symposium*, Vol. 2, pp. 739–742.

[47] Kim, K., Akbar, I. A., Bae, K. K., Um, J. S., Spooner, C. M., & Reed, J. H. (2007). Cyclostationary approaches to signal detection and classification in cognitive radio. In *New Frontiers in 2nd IEEE International Symposium on Dynamic Spectrum Access Networks*, Dublin, Ireland, pp. 212–215.

[48] Chu, Y., Mitchell, P. D., & Grace, D. (2012). ALOHA and q-learning based medium access control for wireless sensor networks. In *2012 International Symposium on Wireless Communication Systems* (ISWCS), Paris, France, pp. 511–515.

[49] Liu, Z., & Elhanany, I. (2006). RL-MAC: A reinforcement learning based MAC protocol for wireless sensor networks. *International Journal of Sensor Networks*, 1(3–4), 117–124.

[50] Raj, A. B., Ramesh, M. V., Kulkarni, R. V., & Hemalatha, T. (2012). Security enhancement in wireless sensor networks using machine learning. In *IEEE 14th International Conference on High Performance Computing and Communication & 2012 IEEE 9th International Conference on Embedded Software and Systems (HPCC-ICESS)*, pp. 1264–1269.

[51] Collotta, M., Cascio, A. L., Pau, G., & Scatá, G. (2013). A fuzzy controller to improve CSMA/CA performance in IEEE 802.15.4 industrial wireless sensor networks. In *IEEE 18th Conference on IEEE Emerging Technologies & Factory Automation*, Cagliari, Italy.

[52] Kulkarni, R. V., & Venayagamoorthy, G. K. (2009). Neural network based secure media access control protocol for wireless sensor networks. In *International Joint Conference on Neural Networks, IJCNN 2009*, pp. 1680–1687.

[53] Qiao, M., Zhao, H., Wang, S., & Wei, J. (2016). MAC protocol selection based on machine learning in cognitive radio networks. In *19th International Symposium on Wireless Personal Multimedia Communications (WPMC)*, pp. 453–458.

[54] Rewadkar, D., & Doye, D. (2018). Multiobjective autoregressive whale optimization for traffic-aware routing in urban VANET. *IET Information Security*, 12(4), 293–304.

[55] Hu, T., & Fei, Y. (2010). QELAR: A machine-learning-based adaptive routing protocol for energy-efficient and lifetime-extended underwater sensor networks. *IEEE Transactions on Mobile Computing*, 9(6), pp. 796–809.

[56] Yang, J., Zhang, H., Pan, C., & Sun, W. (2013). Learning-based routing approach for direct interactions between wireless sensor network and moving vehicles. In *16th International IEEE Conference on Intelligent Transportation Systems-(ITSC)*, pp. 1971–1976.

[57] Huang, G. B., Zhu, Q. Y., & Siew, C. K. (2006). Extreme learning machine: Theory and applications. *Neurocomputing*, 70(1–3), 489–501.

[58] Fahad, M., Aadil, F., Khan, S., Shah, P. A., Muhammad, K., Lloret, J., & Mehmood, I. (2018). Grey wolf optimization based clustering algorithm for vehicular ad-hoc networks. *Computers & Electrical Engineering*, 70, 853–870.

[59] Kumar, K. S., & Arunkumar, T. (2013). Classification and prediction of routing nodes behavior in MANET using Fuzzy proximity relation and ordering with Bayesian classifier. In *2013 International Conference on Pattern Recognition, Informatics and Mobile Engineering (PRIME)*, Salem, India, pp. 454–460.

[60] Zhao, L., Li, Y., Meng, C., Gong, C., & Tang, X. (2016). A SVM based routing scheme in VANETs. In *16th International Symposium on Communications and Information Technologies (ISCIT)*, Qingdao, China, pp. 380–383.

[61] Al-Dubai, A. Y., Zhao, L., Zomaya, A. Y., & Min, G. (2015). QoS-aware inter-domain multicast for scalable wireless community networks. In *IEEE Transactions on Parallel and Distributed Systems*, 26(11), 3136–3148.

[62] Zhao, L., Al-Dubai, A., Li, X., Chen, G., & Min, G. (2017). A new efficient cross-layer relay node selection model for wireless community mesh networks. *Computers & Electrical Engineering*, 61, 361–372.

[63] Di Caro, G., & Dorigo, M. (1998). AntNet: Distributed stigmergetic control for communications networks. *Journal of Artificial Intelligence Research*, 9, 317–365.

[64] Nunes, B. A. A., Veenstra, K., Ballenthin, W., Lukin, S., & Obraczka, K. (2014). A machine learning framework for TCP round-trip time estimation. *EURASIP Journal on Wireless Communications and Networking*, 2014, 47.

[65] Mirza, M., Sommers, J., Barford, P., & Zhu, X. (2010). A machine learning approach to TCP throughput prediction. *IEEE/ACM Transactions on Networking (TON)*, 18(4), 1026–1039.

[66] ETSI GS MEC-IEG 004 V1.1.1. (2015). The Mobile-Edge Computing (MEC) ETSI Industry Specification Group (ISG).

[67] Han, K., Hwang, A., Lee, J. Y., & Kim, B. C. (2018). Design and performance evaluation of enhanced congestion control algorithm for wireless TCP by using a deep learning. In *International Conference on Electronics, Information, and Communication (ICEIC)*, Honolulu.

[68] Chih-Lin, I., Sun, Q., Liu, Z., Zhang, S., & Han, S. (2017). The Big-Data-driven intelligent wireless network: Architecture, use cases, solutions, and future trends. *IEEE Vehicular Technology Magazine*, 12(4), 20–29.

[69] Lin, Y. C. (2017). An intelligent network sharing system design with multi-path TCP. In *18th International Conference on Parallel and Distributed Computing, Applications and Technologies (PDCAT)*, Taipei, Taiwan, pp. 407–413.

[70] Arianpoo, N., Aydin, I., & Leung, V. C. (2017). Network coding as a performance booster for concurrent multi-path transfer of data in multi-hop wireless networks. *IEEE Transactions on Mobile Computing*, 16(4), 1047–1058.

[71] Biswas, M. I., Parr, G., McClean, S., Morrow, P., & Scotney, B. (2014). SLA-based scheduling of applications for geographically secluded clouds. In *2014 International Conference and Workshop on the Network of the Future (NOF)*, Paris, France, pp. 1–8.

[72] Li, H., Ota, K., & Dong, M. (2018). Learning IoT in edge: Deep learning for the internet of things with edge computing. *IEEE Network*, 32(1), 96–101.

[73] Akbani, R., Korkmaz, T., & Raju, G. V. S. (2008, November). A machine learning based reputation system for defending against malicious node behavior. In *2008 Global Telecommunications Conference*, New Orleans, L., USA.

[74] Wang, M., Cui, Y., Wang, X., Xiao, S., & Jiang, J. (2018). Machine learning for networking: Workflow, advances and opportunities. *IEEE Network*, 32(2), 92–99.

[75] Estellés-Arolas, E. & González-Ladrón-de-Guevara, F. (2012). Towards an integrated crowdsourcing definition. *Journal of Information Science*, 38(2), 189–200.

Chapter 3

Adapting Future Vehicle Technologies for Smart Traffic Control Systems

Dominic J. Hodgkiss* and Vinh Thong Ta[†]
University of Central Lancashire (UCLan),
School of Physical Sciences and Computing (PSC),
Preston, UK
Dhodgkiss2@uclan.ac.uk
[†] *Vtta@uclan.ac.uk*

1 Introduction

Traffic control systems are imperative to the everyday functioning and quality of life for a society. The current methods, such as Sydneg Co-ordinated Adaptive Traffic System (SCATS) [1], Split Cycle Offset Optimization Technique (SCOOT) [2] and InSync [3], provide this solution, but with limited flexibility. With the advances in context-aware technologies and wireless vehicular communication as discussed by Maglaras [4], and the rise of the internet of things (IoT) allowing inexpensive networking of devices, current technologies are becoming rapidly outdated. Some examples of such vehicle technologies have been discussed in recent studies, namely, the social internet of vehicles [5, 6] and wireless sensing technologies [7]. As the smart city landscapes develop, some of these technological advances can be adapted into smart traffic control systems (STCSs), improving the transport efficiency throughout the road network, reducing the levels of traffic congestion and amount of air pollution and improving the quality of life. Although air pollution can be somewhat mitigated with technologies like Stop-Start, Hybrid or Electric, traffic congestion still has a negative effect on the quality of life of the drivers as well as the residents in the affected areas. As has been outlined before by Glaesar [8], reducing traffic congestion remains a crucial goal of these future vehicle technologies.

Addressing the traffic congestion problem, this chapter reviews the existing technologies and future vehicle concepts that may be a good starting point for future studies and for implementing an STCS. It starts by looking at the importance of STCSs, reviewing the existing technologies in use with a focus on the most common and identifying their shortcomings. Afterward, three potential vehicular technologies, vehicle-to-X (V2X) communication [9], vehicle cloud computing (VCC) [6] and vehicle social networks (VSNs) [5], will be reviewed based on the previous works [10, 11], with their applicability in STCSs based on potential efficiency, security and privacy aspects.

We decided to choose these three technologies or concepts because they attracted great attention from both research communities and the industry for their potential role in the smart city landscape.

2　Current Technology

Adaptive traffic control systems (ATCS) are designed to manage a traffic junction effectively. These systems use varied sensing equipments to determine the number of vehicles waiting at a given point and then change the lights for an appropriate length of time to allow the optimal number of vehicles through the junction. As a result, the vehicles' waiting time is minimized and congestion is mitigated. See Table A.1 in Appendix for a comparison of the reviewed ATCSs.

2.1　*Sydney coordinated adaptive traffic system (SCATS)*

SCATS is the most widely implemented ATCS across 25 countries, though it was originally created by the Roads and Traffic Authority (RTA) of New South Wales, Australia in 1970 [12]. It is a real-time traffic control system designed to optimize the traffic flow by synchronizing the signals over a whole city, region or corridor. This is done by having a central management server that can connect to 64 regional controllers, each of which can handle up to 250 intersections.

The regional controllers connect to the local intersection controllers which have access to control the lights, as well as gather vehicle presence information from the coils built into the road, called "Loop Detectors". The local controllers measure the traffic intensity using the loop detectors to determine the degree of traffic saturation over a predetermined time [13].

These data are then sent to the regional controllers which calculate the most effective cycle lengths (from 20 to 240 seconds, the time for each road to wait), splits (which change the importance of the main road over minor ones) and offsets for the intersection lights, which the local controllers enact.

Statistics show that SCATS on average reduces the delays by 20%, reducing stops by 40% and therefore reducing fuel consumption by 12% and emissions by 7%, while also allowing emergency services on the fly control to stop traffic for their arrival [14].

2.2　*Split cycle offset optimization technique (SCOOT)*

A technology developed in the United Kingdom, SCOOT, is similar to SCATS in that it uses green-split and offset calculation to manage traffic in real time. Using a centralized architecture and scheduling algorithm, the system uses the induction loop technology built into the road to detect when vehicles pass over the roads. The system works by having a sensor at the start of the traffic light waiting zone of an intersection and another sensor 50–300 m before this waiting point. The system then uses something called cyclic flow profiles to estimate the number of vehicles that enter the road area roughly every 4 seconds. To minimize stops and delays, a queue model is used; this model optimizes the amount of green for each approach called the "split", the time between adjacent signals or "offset" and the time allowed for all approaches to the intersection or "cycle time".

A hierarchical leveling is used for optimization, using Region, Link, Node and Stage as the different levels. Region determines the cycle length optimization, Link prevents queuing occurrence, Node is for fine adjustments of all parameters and Stage sets limits for the minimum and maximum stage lengths. These optimizations can be turned off or on by the system depending on the histograms of traffic saturation in the zone collected hourly, daily and weekly. These inform future flow predictions to determine how to operate the most efficient flow through the intersection. These records also include data, such as occupancy levels, peak flow hours, and queue length.

Upon introduction, certain areas experienced delay reductions of up to 30% over the conventional fixed-time systems; interestingly, during high peak times like sporting events, delay reductions were as high as 61% [2, 14].

2.3　*InSync*

This system differs from the previous two in that it does not use induction loop technology built into the road, with optimization occurring on the splits and offsets. Instead, InSync uses internet protocol (IP) camera systems installed at the intersections to visually detect the number of vehicles approaching and waiting at given points. This is done by having detection zones with contours drawn across them. By counting how many of these contour segments have vehicles in them, the system can determine how many vehicles are waiting, as well as how long they have been waiting there. This allows the system to quantify the traffic saturation at the intersection as well as provide real-time video feed monitoring of the approaches.

The system also avoids the analog style of light control with cycle length, splits and offsets, instead of using something called a finite state machine. This method contains states the intersection can be in, with some states being adaptive so as to account for the varied scenarios and local optimization, by implementing phasing, sequencing and green-time allocation to do so. The system is also much simpler in hierarchy than SCATS or SCOOT as there are only local and global optimizations, with the global able to override the local at any time.

At the global level, traffic is monitored using something called "platoons" which are routed through traffic corridors by altering the green time of intersections to reduce stop times. The local optimizer handles the phasing and sequencing, leaving the green time to the global optimizer, which leaves the local optimizer to control the delay time and volume of vehicles. This is done by using an algorithm to award each vehicle with a weight of importance; an approach with a higher weighting will be given priority over those with a lower weight or those with no vehicles waiting at all. This weighting can be altered as well for different vehicles, such as buses, trucks, and emergency vehicles[3].

2.4 *Shortcomings*

In this section, we will outline some limitations of the current technologies that future technologies will have a potential to improve.

Emergency Responder: Not all the discussed traditional technologies have an emergency responder control function implemented; this can mean responders take longer to move through junctions. If the functionality is in place but is not used at the correct time, then the timings will not be influenced to be ready for the responder's arrival at the junction, voiding the functionalities' usefulness.

Limited Bandwidth: The transfer of information about the state of the traffic signals is slow, as the decision-making time of the drivers slows the process of moving off from an intersection. "When fully aware of the time and location of the brake signal, drivers can detect a signal and move the foot from accelerator to brake pedal in about 0.70 to 0.75 sec." [15]. This pause added up from numerous drivers, as well as the acceleration time, reduces efficiency significantly. If the state of the traffic signals could be sent directly to vehicle computers, not only would there be a record but also, in the case of self-driven vehicles, the decision-making time would be removed. Unfortunately, there are still few vehicles able to receive this information and fewer are able to act on it; however, vehicle automation is on the rise, so adding this functionality could improve future usage.

Inability to Divert: If there is a traffic collision and traffic is building up already, there are currently limited or only manually activated ways to prevent vehicles from routing the same way. Doing this would require the cooperation of the vehicles as the onboard computer would need to recommend the new route to the driver for them to confirm it; of course, this would not be viable in all scenarios.

Limited Sensing Range: The range at which an intersection can receive data is limited to the range of vision of the camera or induction loops installed at an approach. Having these sensors so close prevents the intersection control from preemptively implementing alterations to the lights to further reduce the unnecessary stops of approaching vehicles.

3 Potential Application of Future Technologies in STCS

3.1 *V2V/V2I/V2P/P2I (V2X) communications*

Introduction: V2X is derived from the IoT concept where varying devices are all connected to a network to share sensing and controlled functionality. Many vehicles are equipped with electronic control units (ECU), sometimes with wireless connectivity for maintenance purposes. Also, most intersections are connected via wired connection to control infrastructure. V2X will utilize this connectivity to facilitate information-sharing with all other end points on the road network. Zheng *et al.* [16] has discussed various other nodes for communication in detail. Also, see Fig. A.1 in the appendix.

Potential Role in Traffic Control: By adapting V2X communications in road traffic control, the goal is to facilitate sensory sharing across the road network [7]. This allows each node in the network, irrespective of whether that's a vehicle, base station or pedestrian, to get a more accurate picture of the traffic. Vehicles may alter their route to avoid road congestion and therefore reduce it for others. Intersection control algorithms will be able to determine the most efficient sequence to allow vehicles and pedestrians to move and to improve safety monitoring to mitigate safety risks as they may occur. Relating to road intersections particularly, this would allow information about the approaching vehicle speed, distance and route to inform more efficient sequencing of the lights, so as to reduce unnecessary stops [17]. The current sensing range of an intersection is quite close, whereas with V2X, the information could be received a few miles in advance, allowing the information to be enacted at the correct time and to affect other intersections in the vicinity to compensate.

Challenges: There are several challenges for this technology that need to be resolved, such as lane identification, whether that's at an intersection or traveling through areas of poor GPS signal, mechanisms for incident detection to ensure an appropriate response is made, route sharing that ensures privacy and integration that does not mean old technology is obsolete and incompatible, but only that it is not as efficient as the latest V2X hardware.

3.2   *Vehicular cloud computing (VCC)*

Introduction: As said previously, many vehicles are equipped with ECUs; however, their resources (i.e., storage, processing power, etc.) go unused for long periods of time each day, i.e., when caught in congestion without use or when parked at work or home. See Fig. A.2 in the appendix.

Potential Role in Traffic Control: VCC will make use of surplus or unused resources by creating cloud networks that allow the exchange of the resources for a reward, such as traffic information, connectivity credit, and reduced service cost from an organization. Similar to solar panels which slowly pay back their installation and maintenance costs to the household, this technology could allow vehicle owners to make use of their vehicles in a monetary way. The first paper to develop the idea was written by Abuelela and Olario [18], who suggested "a group of largely autonomous vehicles whose corporate communication, sensing, computing and physical resources can be organized and dynamically allocated to authorized users". It is likely that the more usable resource available is the CPU of the ECU due to the fact that the storage will not be much larger than the operating system and the space for update files and the maintenance and debugging of logs.

When considering this technology specifically for intersections, there will be a lot more sensory data being fed into the intersection controller; greater processing abilities will be required. To facilitate this, the intersection could distribute the required processing to vehicles approaching and waiting at the intersection. In this way, improving the efficiency of the intersection is the reward for the small amount of CPU usage the intersection would take per vehicle.

Challenges: The major issue is that if there is no incentive for allowing your vehicle to become part of a cloud network, then no one will want to, as if it becomes too currency-oriented, there may be reduced cooperation. Anytime a resource is used to benefit the cloud network, there needs to be a good enough reward in return. If the correct balance is struck, vehicles should mutually share information that is important to other road users, such as location, traffic information, etc. There are several other challenges for the introduction of this technology, including how to dynamically set up an *ad hoc* non-static infrastructure on the fly, how to ensure the resource-to-reward system is fair and reasonable and how to distribute the processing dynamically while taking into consideration the end-point dropout.

3.3 *Vehicle social networks (VSNs)*

Introduction: In VSNs, vehicles traveling in groups can form social groups for sharing information or enabling the passengers to play games together during their journey. Compared to traditional social networks, VSNs maintain only short-term social connections built up on the fly. These *ad hoc* social networks are built up based on different aspects, such as the same destination, same route, interests or goals. VSNs also rely on V2X communications; however, its core concept is based on social connection among vehicles. There are several applications similar to VSNs, such as RoadSpeak [19], CliqueTrip [20], SocialDrive [21], Waze [22] and Social on Roads (SoR) [23], with some allowing passengers to share route information and traffic conditions (e.g., Waze, SocialDrive). See Fig. A.3 in the appendix.

Potential Role in Traffic Control: VSNs have the potential to improve traffic efficiency on the road, when vehicles traveling in a certain area form traffic VSNs (TVSNs) to share traffic information, namely, vehicles on one road can share traffic information with vehicles on other roads. For example, Waze [22] is a GPS- and community-based navigation application which users can use to share traffic conditions on certain road segments with each other. On the architectural level, TVSNs can be centralized, decentralized or hybrid. In centralized TVSNs, only V2I-type communications take place with every communication passing through the service provider (e.g., Waze's servers), who creates, manages and maintains the TVSNs. Decentralized TVSNs are based on V2V-type communications, where the TVSNs are built up and managed by the vehicles which are themselves on the fly. Finally, in hybrid TVSNs, the roadside units (RSUs) are also involved in conjunction with smart traffic lights when relevant real-time traffic information is also shared with them.

Challenges: VSNs open some interesting problems due to the *ad hoc* short-term nature of the social networks, such as data-forwarding incentives of non-member nodes (as vehicles outside a TVSN may not be willing to forward messages to save resources) and trust problems with selfish or malicious nodes sending or forwarding inaccurate information to save resources. Finally, it is impossible to completely separate the human aspect from the applications when considering social networks.

4 Discussion

The primary directive of any vehicle technology should be safety. To achieve this, these technologies need to ensure the security of data to prevent misinformation from causing potential risks and the privacy of users to protect them from attackers. Using wireless communications means the attack vector can be from external sources as well as internal ones. Therefore, security measures need to ensure the protection of all data in the network so that only correct information is sent between the vehicles, RSUs and control systems. Also, it should only allow those who are authorized to access sensitive data. This information could include details about vehicle speed and route, vehicle owner payment methods, passenger-specific details, etc., all of which require different levels of protection based on the scenario of requirement. To this effect, permission levels should be implemented so that only certain information is available to different groups of network users. For example, traffic control systems require vehicle-specific details but do not require passenger or payment details; toll roads require payment details but do not require route or passenger information.

Each scenario differs, so it may be difficult to strike a balance. In cases where further information is needed, a request process should be implemented to vet the user. Acquiring this right to share information may improve efficiency networkwide and could reduce unnecessary stops at varied places. Pseudonyms have been discussed before [24] to obscure the user from their data, allowing only authorized users to be aware of which vehicle it belonged to, maintaining the anonymity of data. Another potential option for obfuscation of sensitive data is implementing varying levels of encryption on the data that a vehicle stores; this way, only those with the correct key can decrypt that data.

Tampering with devices connecting to the network will be the first potential threat as attackers attempt to reverse-engineer hardware and communication protocols to determine vulnerabilities. Two researchers were able to gain control of critical vehicle systems (like brakes, steering, accelerator, etc.) and control them from 10 miles away, connecting via 3G using vulnerabilities in the Uconnect software used by the vehicle [25]. Antitampering techniques (e.g., inductive switches, hall effect sensors) can prevent or detect physical tampering; however, attackers may still find ways to circumvent them. Therefore, reactionary methods should be used to inform of a rogue node as information from it is likely to be incorrect. This is a persistent issue of all technologies discussed as developing trust between network nodes is tricky [26] and this is especially true in the case of technologies like VCC where resources are of monetary value. Implementing the vehicle profile aspect of VSNs into all handshake protocols to determine the other node's trustworthiness could be a potential solution to ensure they are tamper-free, based on previous communications. Current methods of authentication could be used with some alterations, perhaps a twin authentication where each node communicates with an infrastructure node to validate authenticity, or within the firmware, to flag whether the system has been tampered with based on the sensors in the casing.

Of course, breaches will occur; in these scenarios, fail-safes should be implemented to ensure public safety and avoid misinformation from traffic control systems being manipulated into causing serious damage. This applies anywhere on the road network where a set of static rules should always operate, e.g., route 1 stops before route 2 can move. Physical redundancies also ensure any service drops are for as minimal a time as possible; if an RSU or intersection controller goes down, there should be another on standby ready to come online until repairs are made. Considering the security implications of a downed station, all vehicles in the vicinity will be looking to connect to a station, which leaves a hole in the network that an attacker could fill with a rogue station.

Unfortunately, even fail-safes and redundancies will sometimes not be enough to avoid incidents, such as when traffic lights have been misled by fake or incorrect information coming from vehicles or RSUs. In these scenarios, there should be procedures in place to ensure that any incident is reported and investigated correctly. Though this will be heavily reliant on the scenario, each vehicle will likely be logging a lot of information, like a black box in an airplane. This information should contain a histography of speed, intended route, steering wheel angle, etc. This information could then be used to determine all actions from the involved vehicles up to the point of the incident to allow an investigator to understand how and why an incident occurred, as well as who is at fault. This also relates to physical infrastructure nodes, as if there is a service drop, the risk of an incident occurring may increase. In these cases, an investigation would need to be conducted to find out whether it was a lack of maintenance, energy brownouts or surges, incorrect manipulation by the regulator, etc. One of the most important things to ensure investigations can be conducted correctly is the existence of extensive logs storing as much data as possible on all actions taken by the road users, system checks and network traffic.

To ensure that all manufacturers integrating these technologies into their products are compatible with each other, a set of standards would need to be adhered to. Within the IEEE, there is a technical subcommittee that regularly reviews Vehicular Network & Telematics Applications (VNTA). Since VCC and VSNs are in their infancy, mainly at the research stage, no standardization can be found in these areas; there are however several international standards related to the V2X concept such as the IEEE 1609 Wireless Access in Vehicular Environments (WAVE), which is built on 802.11p WLAN and designed to add multichannel operation, security and a lightweight application layer [27].

It is important to consider future technologies like machine learning (ML) as a potential resource for managing and improving efficiency and safety, while allowing persistent monitoring and control, potentially identifying risks and preventing them before they become serious. Emergency responses would be quicker as detection and alerting would be instant. Intersections would also be controlled to assist in the fast transit of emergency responders. The intelligence would be instantly aware of downed stations and could inform vehicles to route packets through each other to the closest stations to avoid drops in service, as well as to send a maintenance and investigation request to the appropriate individuals.

5 Conclusion

In this chapter, we discussed possible future STCSs based on three future vehicular technologies that have the potential to overcome the limitations posed by the current, widely used, ATCSs, with an overview of the current traffic control systems and their limitations, as well as highlighting future vehicular concepts, such as V2X communications, VCC and VSNs. We discussed the possible application of these three concepts for smarter and more efficient traffic control. The challenges and problems of each technology were raised to inform possible research directions. We then discussed the common problems V2X, VCC and VSNs may suffer related to effectiveness, security and privacy, such as trust and selfishness problems, tampering issues and the complications involved in investigating incidents. Lastly, we have mentioned other vehicular future technologies that could be examined in this context; however, due to the space limitation, we have decided to focus on these three concepts as they are the most interesting and actively researched areas.

Appendix

Table A.1. ATCS comparison.

	SCATs	**SCOOT**	**InSync**
Place of deployment	Australia/Oceania, Asia	UK, Commonwealth	USA
Place of design	Australia	UK	USA
Traffic intensity is measured with	One set of inductive loop detectors, in-road sensors	Two sets of inductive loop detectors	IP camcorders
Traffic intensity is measured at	Stop line/every lane	Stop line and upstream end of the road/every lane	Stop line/every lane
What is measured?	Distance between vehicles	Saturation, one-way flow profile	Image processing, vehicle counting
Traffic schedule/ optimization is based on	Cycle lengths, splits and offsets	Cycle lengths, splits and offsets	Finite state machine states of the intersection
Architecture	Centralized/ hierarchical	Centralized/ hierarchical	Centralized/ hierarchical

 D. J. Hodgkiss & V. T. Ta

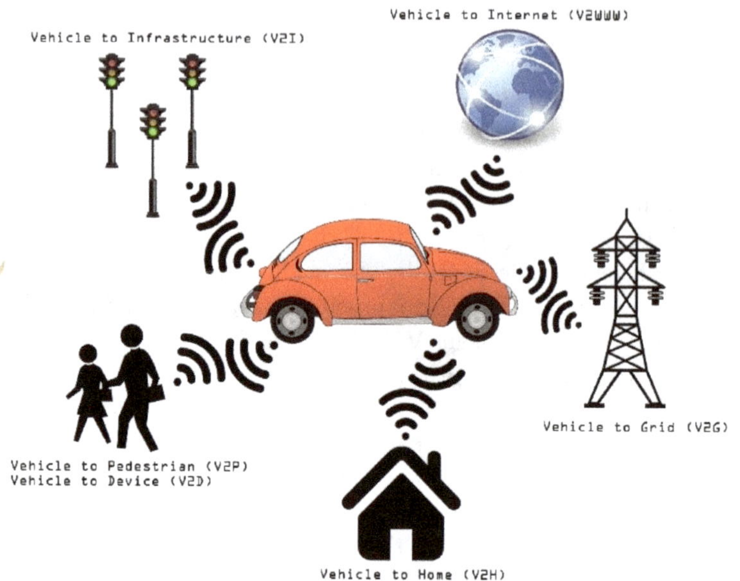

Fig. A.1.　V2X communications.

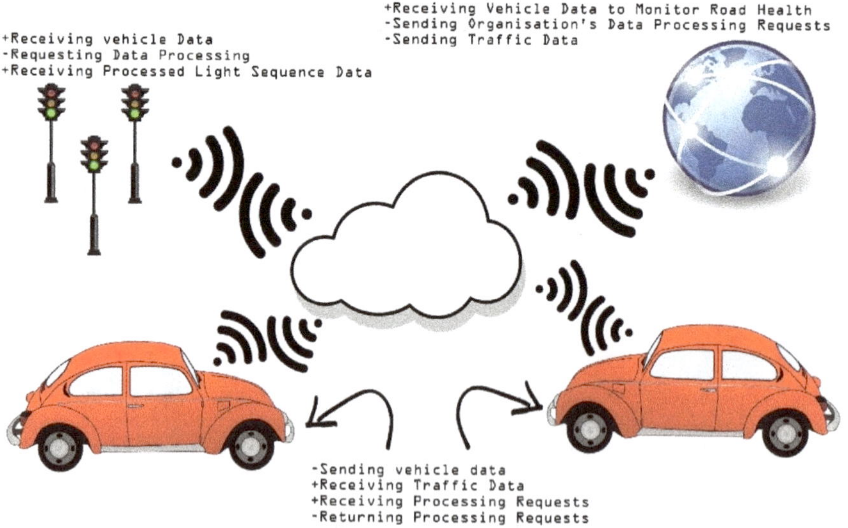

Fig. A.2.　VCC concept in smart traffic control.

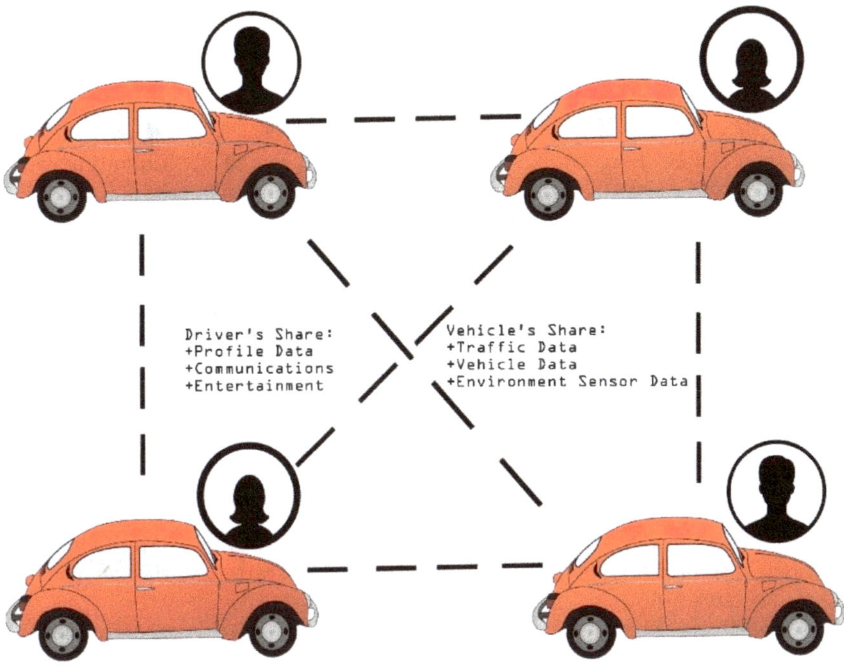

Fig. A.3. VSN concept in smart traffic control.

References

[1] SCATS [Online]. http://www.scats.com.au/ [Accessed 4th May 2018].

[2] SCOOT, TRL Software [Online]. https://trlsoftware.co.uk/products/traffic_control/scoot [Accessed 4th May 2018].

[3] InSync Rhythm Engineering [Online]. https://rhythmtraffic.com/insync-2/ [Accessed 4th May 2018].

[4] Maglaras, L. A., Maglaras, L. A., He, Y., Wagner I., & Janicke, H. (2016). Social internet of vehicles for smart cities. *Journal of Sensor and Actuator Networks*, 5(1), 3.

[5] George, S. P., Wilson, N., Nair, K. U., Michael, K., & Aricatt, M. B. (2017). Social internet of vehicles. *International Research Journal of Engineering and Technology*, 4(4), 712–717.

[6] Gerla, M., Lee, E.-K., Pau, G., & Lee, U. (2014). Internet of vehicles: From intelligent grid to autonomous cars and vehicular cloud. In *IEEE World Forum on Internet of Things (WF-IoT)*, Piscataway, pp. 241–246.

[7] Bhuvaneswari, P. T. V., Raj, G. V. A., Balaji, R., & Kanagasabai, S. (2012). Adaptive traffic signal flow control using wireless sensor networks. In *Fourth International Conference on Computational Intelligence and Communication Networks (CICN)*, Mathura, India, pp. 85–89.

[8] Glaesar, E. (2011). Cities, productivity and quality of life. *Science*, 333(6042), 592–594.

[9] Bilstrup, K., Uhlemann, E., Strom, E. G., & Bilstrup, U. (2008). Evaluation of the IEEE 802.11p MAC method for vehicle-to-vehicle communication. In *IEEE 68th Vehicular Technology Conference, VTC 2008-Fall*, pp. 1–5.

[10] Sinai, M. B., Partush, N., Yadid, S., & Yahav, E. (2014). Exploiting social navigation. Cornell University Library, arXiv.org, New York.

[11] Papageorgiou, M., Diakaki, C., Dinopoulou, V., Kotsialos, A., & Wang, Y. (2003). Review of road traffic control strategies. *Proceedings of the IEEE*, 91(12), 2043–2067.

[12] Zhao, Y., & Tian, Z. (2012). An overview of the usage of adaptive signal control system in the United States of America. *Applied Mechanics and Materials*, (178–181), 2591–2598.

[13] Samadi, S., Rad, A. P., Kazemi F. M., & Jafarian, H. (2012). Performance evaluation of intelligent adaptive traffic control systems: A case study. *Journal of Transportation Technologies*, 2, 248–259.

[14] Ta, V. T. (2016). Automated road traffic congestion detection and alarm systems: Incorporating V2I communications into ATCSS. *CoRR*.

[15] Green, M. (2000). How long does it take to stop? (2000). Methodological analysis of driver perception-brake times. *Transportation Human Factors*, 2(3), 195–216.

[16] Zheng, K., Zheng, Q., Yang, H., Zhao, L., Hou, L., & Chatzimisios, P. (2015). Reliable and efficient autonomous driving: the need for heterogeneous vehicular networks. In *IEEE Communications Magazine*, 53(12), 72–79.

[17] Obst, M., Hobert, L., & Reisdorf, P. (2014). Multi-sensor data fusion for checking plausibility of V2V communications by vision-based multiple-object tracking. In *IEEE Vehicular Networking Conference*, pp. 143–150.

[18] Abuelela, M., & Olariu, S. (2010). Taking VANET to the clouds. In *Proceedings of the 8th International Conference on Advances in Mobile Computing and Multimedia*, pp. 6–13.

[19] Smaldone, S., Han, L., Shankar, P., & Iftode, L. (2008). RoadSpeak: Enabling voice chat on roadways using vehicular social networks. In *the 1st Workshop on Social Network Systems (SocialNets '08)*, New York.

[20] Knobel, M., Hassenzahl, M., Lamara, M., Sattler, T., Schumann, J., Eckoldt, K., & Butz, A. (2012). Clique trip: Feeling related in different cars. In *Proceedings of the Designing Interactive Systems Conference (DIS '12)*, New York.

[21] Xiping, H., Victor, L. C. M., Kevin, G. L., Edmond, K., Haochen, Z., Nambiar, S. S., & Peyman, T. (2013). Social drive: A crowdsourcing-based vehicular social networking system for green transportation. In *Proceedings of the Third ACM International Symposium on Design and Analysis of Intelligent Vehicular Networks and Applications (DIVANet '13)*, New York.

[22] Waze, Waze — Free community-based mapping, Traffic & Navigation App [Online]. https://www.waze.com/ [Accessed 28 September 2018].

[23] Luan, T., Lu, R., Shen X., & Bai F. (2015). Social on the road: Enabling secure and efficient social networking on highways. *IEEE Wireless Communications*, 22(1), 44–51.

[24] Bjorn, W., Sall, M., & Reinhard, G. (2009). SeVeCom — Security and privacy in Car2Car ad hoc networks. In *International Conference on Intelligent Transport Systems Telecommunications, (ITST)*, Lille.

[25] Greenberg, A. Hackers remotely kill a Jeep on the highway — With me in it. *WIRED*, 21st July 2015 [Online]. https://www.wired.com/2015/07/hackers-remotely-kill-jeep-highway/.

[26] Wex, P., Breuer, J., Held, A., Leinmuller, T., & Delgrossi, L. (2008). Trust issues for vehicular ad hoc networks. In *IEEE Vehicular Technology Conference*, Singapore.

[27] U. D. O. Transportation (2013). IEEE 1609 — Family of standards for wireless access in vehicular environments (WAVE).

Chapter 4

Application of Mobile Signaling in Highway Traffic Network Monitoring

Hongrun Gang[*,‡], Li Fu[*,§] and Yuting Luan[†,¶]

*China Academy of Transportation Sciences,
Beijing, P. R. China
†China Railway Engineering Consulting Group Corporation,
Beijing, P. R. China
‡ganghongrun@163.com
§li.fu@live.cn
¶176976816@qq.com

1 Introduction

1.1 *Research status*

Traffic data analysis based on mobile signaling is a low-cost and highly popular technology. This technology has been widely researched by both academics and industries.

The earliest use of the mobile phone as a location application dates back to the 1990s. Initially, a USA project used the triangular relationship between mobile signaling and base station location to analyze and calculate the speed of motor vehicles, which were widely used in military programs at that time [1]. Since 1999, some research institutes have begun to use business data in mobile communication networks for traffic flow analysis. In 1999, White and Well used mobile phone bill data to analyze the population flow [2]. In 2000, Yim and Cayford used 44 hours of wireless data, including the location of the mobile phone during a voice call, to

65

analyze the traffic flow [3]. In 2001, Smith *et al.* designed and implemented an experiment to calculate the traffic speed using a relatively large amount of mobile phone data, which used 160 special mobile phones. These mobile phones reported their location data every 2 seconds (such that 4,800 points of data were recorded per minute). Traffic speed analysis was performed successfully using these location data and base station data [4]. In 2001 and 2002, SRTIP and Kummala used the mobile phone call data and the cellular base station switching data to track the travel routes of the mobile user as well as both the traffic conditions and the traffic speed [5, 6]. After 2003, with the further enhancement of data management capability and the improvement of streaming data processing capability, mobile signaling data can be collected and managed. Furthermore, mobile communication network logs can be analyzed, which is mainly based on analyzing the HandOver data of the base station switching, the heartbeat and UpLocation data of the regional switch and the Erlang data of base station pressure. In 2003, Thiessenhusen *et al.* used the HandOver data of an anonymous phone to carry on the evaluation and analysis of the traffic flow data [7]. In 2004 and 2005, Rutten and UVCTS further used the HandOver and UpLocation data to analyze the traffic flow and traffic jam [8, 9]. In view of the traffic flow problems of specific travel routes, Maerivoet and Logghe [10] and Bar-Gera [11] analyzed the traffic flow of some travel routes using Erlang data in 2006. In 2005 and 2006, MIT Senseable [12] and RealRoma [13] projects used Erlang data to analyze the traffic conditions of the whole city, such as call density, origin–destination (OD) analysis of traffic pedestrian distribution and traffic pressure.

Since 2007, the development of IOT technology has led to a combination of high-precision data collection and traditional Big Data analysis. The development of Big Data storage and analysis technology provides technical support for super large-scale data analysis. The research work of modeling the whole traffic state and analyzing the whole spatial analysis system is also emerging progressively.

In 2014, researchers in Belgium analyzed the Portuguese and French migration populations between July and August 2014 based on mobile phone signaling data. Furthermore, they found that the population in major cities decreased during the 2 months, while the population in coastal resort areas increased [14–16]. In addition, the experimental results were also compared with the remote sensing data, which proved the correctness of the analysis.

1.2 *Application case*

With the enhancement of Big Data computing and analysis capability, mobile signaling data-based large-scale mining applications have been launched worldwide. In the area of traffic condition monitoring and application, governments have been actively exploring the applications of mobile signaling data based on the demands of traffic management. Meanwhile, telecom operators have integrated and developed many products on the basis of existing signaling data. In addition, many commercial companies began to aggregate telecom operators' mobile signaling data to explore transportation-related applications.

In 2005, the state government of Missouri began to use the driver's mobile phone data to collect road operation information. The speed of vehicles can be computed based on these data, and then messages warning of traffic jams would be sent to the driver in time. In addition, the electronic route map and traffic website information would be updated automatically; furthermore, text messages that reflect the traffic conditions would be sent to the user's phone and vehicle terminal. The Missouri department of transportation used the mobile phone signaling data of a large number of users to monitor the road traffic, which used mobile phone movements to reflect real-time traffic conditions on the state's 5,500 miles of main roads. The Missouri program tracked phone signaling data across the whole state, including the rural areas, where traffic is relatively smooth, to provide clear traffic information to drivers [17].

As France's largest telecom operator, Orange has undertaken a French highway data monitoring project. This project produces 5 million records per day, and the analysis of these records can provide accurate and timely expressway operating information for drivers, which effectively improves the road patency rate [18]. In 2012, Telefonica, which is the top telecom operator in Spain, established the Dynamic Insights Department to carry out Big Data business, providing data analysis and packaging services for customers. Dynamic Insights, in collaboration with Gfk Company, a market research group, launched its first product, Smart Steps, in the U.K. (see Fig. 1) and Brazil in 2014. Based on mobile signaling data, this product analyzes the key influencing factors of traffic flow data in a certain period and a certain place and then provides the results to government and enterprise customers [19].

AirSage and Streetlight Data (see Fig. 2) are typical signaling data resource operators. AirSage can collect, anonymize, encrypt and analyze

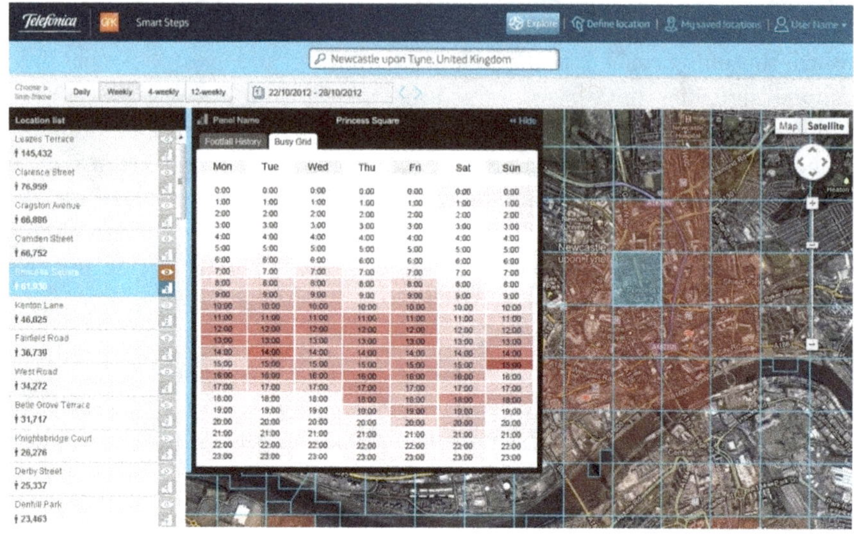

Fig. 1.   The Smart Steps product in the U.K.

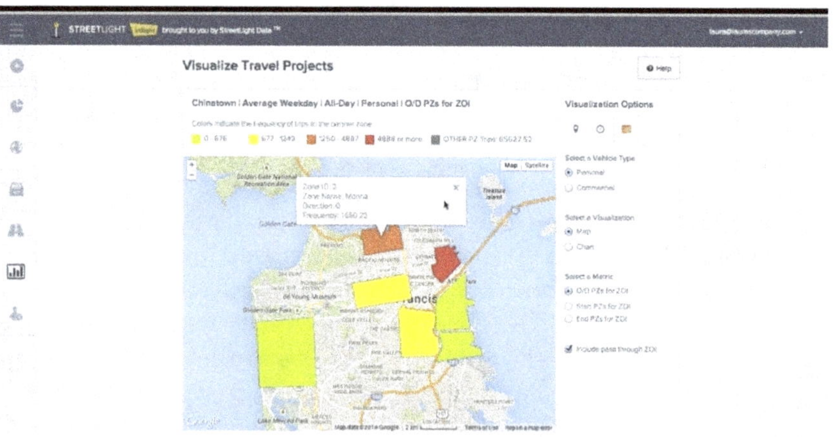

Fig. 2.   Streetlight's personnel flow visualization project.

mobile signaling data in real time from three major U.S. mobile operators. AirSage processes 15 billion points of location data everyday. So far, AirSage's product has been able to translate real-time mobile phone signals into meaningful location, movement and traffic data, and AirSage has made some applications to improve traffic planning and traffic reporting based on that data. In 2009, WiSET, a transportation service product launched by AirSage, was integrated with Google Maps to provide information on the status of traffic in more than 20 markets [20]. By purchasing mobile phone signaling data, Streetlight Data conducted U.S. household census and regional transportation surveys for the government, which are of great significance for analyzing the relationship between traffic behavior and economic development. Economic layout, urban planning and transportation planning in neighboring areas are supported by the data of regional personnel movement. In terms of road operation monitoring, Streetlight conducted a case study involving the road network operation monitoring of the San Diego south bay highway. The San Diego government bought the adjacent highway so that they could divert another highway to ease congestion. A year later, the San Diego government needed to show that the invested roads had addressed the heavy congestion problem, and Streetlight offered a before-and-after comparison, analyzing months of data on both roads before and after tolling. Streetlight's report clearly pointed to the changes in driver behavior regarding geography (e.g., trips in the north and south were compared) and time (e.g., weekday and weekend traveling times were compared), demonstrating the effectiveness of government decision-making [21].

2   China Highway Network Monitoring Background

2.1   *The introduction of highways in China*

Since 2000, the development of China's highway transportation infrastructure has undergone tremendous changes. The scale of the highway network has increased from 1.698 million km in 2001 to 4.7735 million km in 2017 [22], realizing leapfrog development and releasing the potential of China's road transportation network. According to the National Highway Plan [23], the total size of the road network in mainland China will be about 5.8 million km by 2030, of which the national highway will be about 400,000 km (see Fig. 3). At the same time, with the fast development of economy and society, the demand for highway transportation is growing rapidly.

By end of 2017, the number of motor vehicles in China had reached 217 million, and the number of vehicle drivers had exceeded 342 million [23]. With the gradual perfection of the road network and the rapid growth of the demand for road transportation, the pressure of the national road network

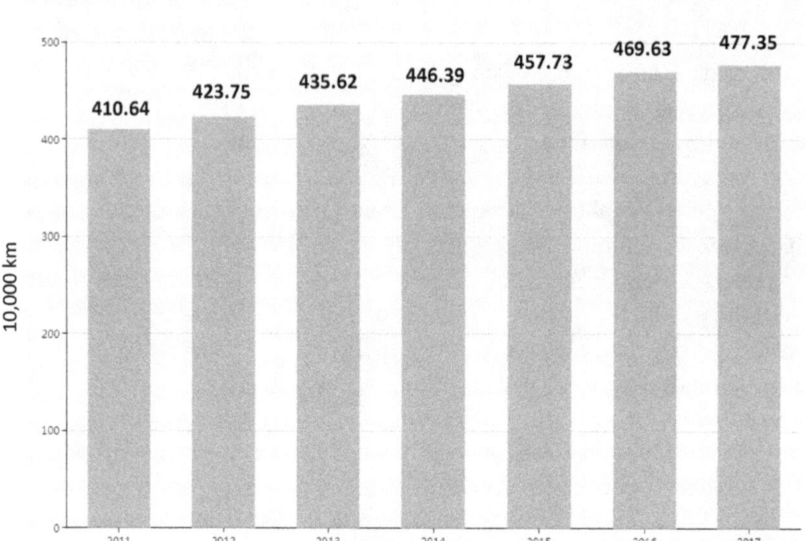

Fig. 3. Total length of national highway from 2011 to 2017.

and the intensity of transportation are further increased. People's travel requirements, i.e. safety, convenience and comfort, cannot be achieved with only infrastructure improvements.

2.2 *China highway network monitoring introduction*

By the end of 2017, the national trunk road network operation monitoring system mainly included the emergency information manual reporting system and the network operation information automatic acquisition system. The emergency information manual reporting system, which is web-based, basically covers all the sections of the national highways and provincial trunk highways, as well as the provincial and municipal road network management departments, and is an important data source of the travel information service system.

According to the incomplete data statistics reported by various places, the total scale of China's highway traffic flow monitoring facilities reaches nearly 10,000 sets, and the average layout density is 25–30 km/set. Video monitoring facilities have a total scale of nearly 25,000 sets, with an average density of 8–10 km/set. The total scale of meteorological monitoring facilities is nearly 1,500 sets, with an average density of 80–100 km/set. The total scale of China's traffic flow monitoring facilities for ordinary national and provincial roads reached nearly 4,000 sets, with an average density of 200 km/set. Video monitoring facilities have a total scale of nearly 2,500 sets, with an average density of 340 km/set. The total scale of meteorological monitoring facilities is nearly 100 sets. In addition, highway toll plazas, super-large bridges and long tunnels are basically covered with traffic flow and video image monitoring facilities.

On the whole, the monitoring facilities (see Fig. 4) of China's arterial highway network have made remarkable progress in terms of construction scale and quality in the past decade. The operation information automatic acquisition system constructed for arterial highways is obviously superior to that of the national and provincial highways. However, there are still problems, such as the small-scale, uneven regional distribution and the large quality gap.

In recent years, with the popularity of smart phones, location services based on mobile communication have become a new wide-area dynamic

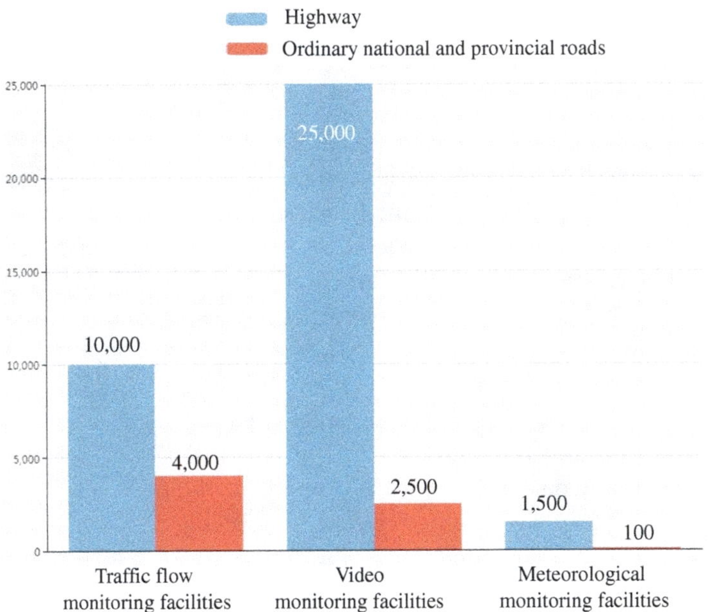

Fig. 4.   Monitoring facilities in China.

traffic detection method. It is an important application of "Internet+" strategy in road network management to analyze and infer the dynamic traffic conditions through the location of mobile phone users. Compared with the traditional network monitoring method, the traffic detection method is based on the existing mobile communication network resources to carry out the traffic information acquisition and analyze the real-time road network running status. The traffic detection method has a series of advantages, such as wide coverage, high real-time performance, short deployment time, low maintenance cost and great reliability.

3 Signaling Network Real-Time Monitoring Platform

A signaling network real-time monitoring platform can extract and filter the original signaling data from the mobile operators' signaling interface and the average speed of each highway can be calculated by computing models, which represent the road network's overall running status. It would be more accurate to calculate values in combination with video image data, vehicle inspection data, accident event data, etc. Signaling data should be processed by the mobile communication operators to obtain traffic congestion information, which can be released to users without any disclosure of users' privacy.

3.1 *Signaling acquisition criteria*

A signaling network real-time monitoring platform uses smart phones to turn vehicles into intelligent vehicles and hence, the running speed and path of the vehicle can be obtained by analyzing the user's mobile phone signaling data. The platform extracts the Global System For Mobile Communications (GSM) signal and the General Packet Radio Service (GPRS) business of the terminal in real time from the network interface of mobile operators (A and GB interfaces). The data types collected includes voice, SMS, websites visited, switches made, etc. The base station within 3 km on both sides of the road network is used as the target area for signal collection. The signaling format is given in Table 1.

Experimental data show that the proportion of interference and invalid signaling is high in the real-time signal that is collected. Meanwhile, the location information contained in mobile signaling belongs to the cell localization based on the location of the base station, which is not extremely accurate, and the distribution is uneven. Therefore, it is necessary to establish the corresponding road model, develop a method to screen the signal and analyze rules to obtain more real and available data.

Table 1. The signaling format.

Name	Description
USERID	IMSI MD5 encrypted string
STARTTIME	YYYY-MM-DD HH:MM:SS.1234567. Accurate to 0.1 microseconds, the signaling start time.
SDRTYPE	11: MO Calling, 12: MT Calling, 31: ON/OFF, 32: Location updated, 33: Handover, 51: SMS MO, 52: SMS MT
CGI	Cell global identifier e.g., 460005123501234
LONGITUDE	LONGITUDE
LATITUDE	LATITUDE
RANTYPE	1 = 2G 2 = 3G 3 = 4G Unknown: 0

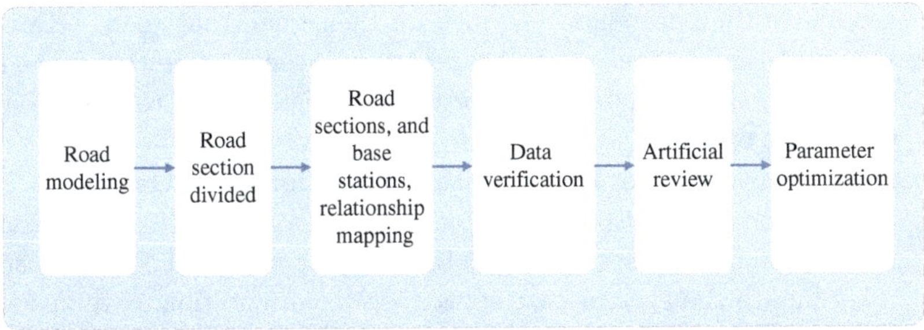

Fig. 5. The main modeling processes of the road monitoring model.

3.2 *Road monitoring model*

In this part, the road monitoring model is established based on the data from the vehicle (mobile phone), road, base station and traffic flow. The model will be affected by factors, such as road direction, road speed limit, historical traffic flow, base station distribution and the surrounding environment.

The main modeling processes of the road monitoring model (see Fig. 5) include road modeling, road section division, road section and base station relationship mapping, data verification, manual auditing and parameter tuning.

(1) **Road Modeling:** Through the highway track drawing system, the starting and stopping points of the route as well as the crossing points are input to generate the route, which can be manually adjusted. If

the route is wrong, it could be that the starting point did not land properly on the road; then the next step is to zoom in to make sure it lands in the right lane. Check whether the route is on the road and if the direction of the route is correct, then automatically generate the road coordinate sequence after route confirmation, and finally import the road information into the database. The road information includes the road code, road name, province code, province name, road speed limit, coordinate sequence, etc.

(2) **Road Section Division:** According to the distribution of the base stations and the extent of coverage of the road by the base station signal, the base station is superimposed on the road and the range of its signal is analyzed as the buffer radius with the use of the GIS buffer analysis technology. The road is divided into several logical sections according to where the base stations are. The strength of the base station signal is also recorded in the model. If a road section is close enough to the base station to pick up the signal, that road section is within the online band of the base station. If the road section is too far away from the base station to pick up the signal, that road section is within the offline band of the base station.

(3) **Road Section and Base Station Relationship Mapping:** The base station information data, after screening and road test calibration, is imported into the system. Then the mapping parameters are adjusted according to the type of base station, sector orientation, coverage and vertical distance from the road section. According to the superposition analysis of the base station and the road, the road base station sequence is obtained as the base station switching sequence of the road section. The mapping parameters are set to generate the mapping relationship between the road sections and the base stations.

(4) **Data Verification:** Based on the historical signaling data, the road monitoring model of the road section and the base station is verified, including the critical path base station, base station signaling, road section signaling, etc. These results of the analysis are verified against the historical data.

(5) **Artificial Review:** The content of road modeling artificial review includes the number of base stations, the base station signaling volume, the total signaling amount of road sections and other indicators. If it fails to meet the application requirements, one has to adjust the configuration parameters of the road monitoring model and redraw the mapping relationship between the road sections and the base stations. Only the approved road monitoring model can be used officially.

(6) **Parameter Optimization:** Parameter optimization is carried out for the road sections that do not meet the requirements, including section range, both side ranges of the road, road section division parameters, base station calibration parameters, etc. The mapping relationship between the road sections and the base stations is redivided, and the road monitoring model is generated. Data verification and review are conducted in the cycle until the road monitoring model meets the application requirements.

3.3　*Data analysis and fusion*

First, the signaling data generated along the highway is obtained from the original signaling data of the mobile operators' interface, screened by combining the base station data of the mobile communication network along with the highway and the filtering rules of the system configuration (see Fig. 6). Then, based on the predefined format, the metadata is output for analysis of the road network operation status according to the monitored highway grouping.

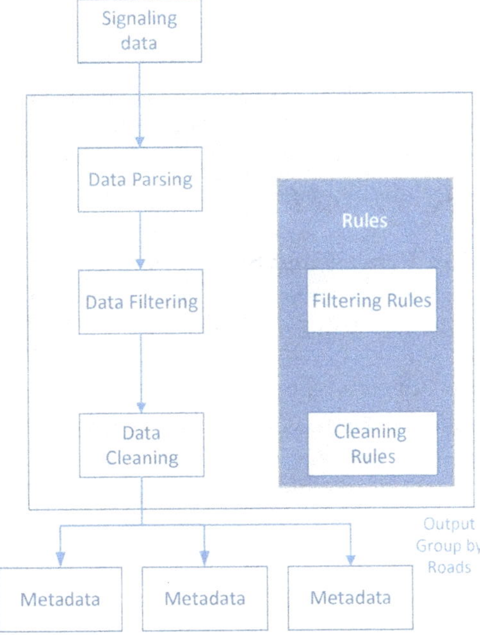

Fig. 6.　Data analysis and fusion.

For signaling data filtering, the input original data are preprocessed and then parsed and filtered according to predefined fields. The following signaling data irrelevant to network signaling should be filtered out:

(1) **Resident User Signaling:** The signaling always appears in adjacent base station areas.
(2) **Ping-pong Switching Signaling:** The switching signaling is repeated in two or three base station areas.
(3) **Abnormal Signaling:** The signaling is not consistent.
(4) **Special Location Signaling:** The signaling occurs in on–off ramps and service areas.

For signaling data cleaning, the signaling data should be further cleaned according to predefined rules. The following are the main cleaning types:

(1) **Incorrect Signaling Format Processing:** The signaling data are discarded if the signaling event is unrecognized.
(2) **Signaling Delay Processing:** The signaling data are discarded if the signaling is delayed too long according to the time stamp field.
(3) **Signaling Group Processing:** The signaling data are discarded when its Location Area Code/Cell Identity (LAC/CI) field cannot be searched for in the base station database.
(4) **User Feature Processing:** The signaling data are discarded if the user identity field in signaling events can be searched for in the relevant database generated by the system learning.
(5) **User Event Correlation Analysis Processing:** The signal is regarded as an abnormal signal and discarded if the user's historical events are abnormal.
(6) **Customization Processing:** Several rules can be defined in advance in combination with the experience of operation, such as the use of a black or white list mechanism.

After grouping, screening and filtering, the signaling data will be output in the predefined format to the next phase for further calculation.

Based on the event information contained in the signaling data (e.g., user id, time, base station cell), the platform can restore the track depicting the movement of individual users between base station areas and update the status of the user in real time according to the user's movement track. According to the road characteristics and user characteristics data, the sample data conforming to real-time road condition analysis is extracted, and the speed of each segment of the user's track within the current statistical window is calculated, according to the geographical location associated with user signaling events and the time difference between the adjacent events. We can calculate the speed of each track of the user based on the following formula:

$$\text{Model speed} = \text{distance/time difference} \left(V = \frac{\Delta L}{\Delta T} \right).$$

Then the platform analyzes the real-time traffic based on a series of relevant data, including the tracks and speed of the multiple users, user characteristics, road characteristics and other preset conditions. After the data filtering and effective sample statistics, analysis and calculation, as well as the transportation data aggregation analysis (vehicle inspection data, traffic incident data, etc.), the monitored road operation status data would be calculated comprehensively.

4 Conclusion

Mobile phone signal data analysis can provide us a new solution to rapidly obtain vehicle operation status on the road network by treating mobile phones as representatives of vehicles. It improves automatic monitoring on the road network and can expand monitoring coverage to the whole road network. It can provide comprehensive, accurate and real-time road network operation information to support upper level road network synthetic management and guarantee the implementation of "situation awareness" on the whole road network.

The platform introduced in Section 3.3 also provides the public with safe, smooth, convenient and environmentally friendly road travel services through various convenient and quick information release methods (e.g., induction screen, radio, text message, microblog) to improve the travel information service level.

References

[1] University of Maryland Transportation on Studies Center. Final evaluation report for the CAPITAL-ITS operational test and demonstration program. University of Maryland, College Park, USA (1997).

[2] White, J., & Well, I. (2002). Extracting origin destination information from mobile phone data. In *Proceedings from 11th International Conference on Road Transport Information and Control*, Vol. 486, pp. 30–34.

[3] Yim, Y. B. Y., & Cayford, R. (2001). Investigation of vehicles as probes using global positioning system and cellular phone tracking: Field operational

test. Report UCB-ITSPWP-2001-9. California PATH Program, Institute of Transportation Studies, University of California, Berkeley.

[4] Smith, B. L., Pack, M. L., Lovell, D. J., & Sermons, M. W. (2001). Transportation management applications of anonymous mobile call sampling. In *TRB 80th Annual Meeting Compendium of Papers CD-ROM.*

[5] Ygnace, J. L. (2001). Travel time/speed estimates on the French Rhone corridor network using cellular phones as probes. Final report of the SERTI V program, INRETS, Lyon, France.

[6] Kummala, J. (2002). Travel time service utilizing mobile phones, Finnish Road Administration Report, Helsinki, Vol. 55, p. 67.

[7] Thiessenhusen, K. U., Schafer, R. P., & Land, T. (2003). Traffic data from cell phones: A comparison with loops and probe vehicle data. Institute of Transport Research, German Aerospace Center, Germany.

[8] Rutten, B., Vlist, M., & Wolff, P. (2009). GSM as a source for traffic information. In *Proceedings of European Transport Conference*, July 29, Strasbourg, France.

[9] University of Virginia Center for Transportation Studies. Wireless Location technology-based traffic monitoring demonstration and evaluation project. Charlottesville, VA (2006).

[10] Maerivoet, S., & Logghe, S. (2007). Validation of travel times based on cellular floating vehicle data. In *proceedings from 6th European Congress and Exhibition on Intelligent Transport Systems and Services.* Aalborg, Denmark.

[11] Bar-Gera, H. (2007). Evaluation of a cellular phone-based system for measurements of traffic speeds and travel times: A case study from Israel. *Transportation Research Part C: Emerging Technologies*, 15(6), 380–391.

[12] Calabrese, F., Reades, J., & Ratti, C. (2010). Eigenplaces: Segmentin space through digital signatures. *Pervasive Computing*, 9(1), 78–84.

[13] Calabrese, F., & Ratti, C. (2006). Real time Rome. *Networks and Communication Studies*, 20(3–4), 247–258.

[14] Deville, P., Linard, C., & Martin, S. (2014). Dynamic population mapping using mobile phone data. *PNAS*, 111(45), 15888–15893.

[15] Burke, J. E. (1964). Scattering of surface waves on an infinitely deep fluid. *Journal of Mathematical Physics*, 6, 805–819.

[16] Fei, Y. (2013). Link travel speed data capture technology based on cellular handoff information: Method, algorithm and evaluation, Science Press, Beijing.

[17] Shu, W. Q. (2012). France telecom: Embrace big data industry opportunities. *Communications World Weekly*, 43, p. 13.

[18] Jie, S. (2012). Mining data and information business potential Spain Telecom and Verizon set up big data Department. *Communications World Weekly*, 43, p. 12.

[19] AirSage (2013). The future of transportation studies: A comparative review [EB/OL]. www.airsage.com.

[20] Leber, J. (2013). How can mobile operators make money with your data. *Technology Review*, 07, pp. 14–18.

[21] Statistical communication of national toll roads in 2017. www.mot.gov.cn. 2018.

[22] National Highway Plan (2013–2030). www.mot.gov.cn. 2018.

[23] Vehicle ownership in China. www.mps.gov.cn. 2017.

Chapter 5

Passenger Trajectory Generation Using Modern Datasets

Xiaoyan Xie[*] and Fabien Leurent[†]
*LVMT, UMR-T 9403, Ecole des Ponts,
IFSTTAR, UPEM, UPE, 77455, Champs-sur-Marne, France*
*xiaoyan.xie@enpc.fr
†fabien.leurent@enpc.fr*

1 Introduction

The smart city provides smart technologies to remedy several urban issues in cities across the world, such as traffic congestion, air pollution, energy consumption, climate change and waste production, which are caused by the rapid growth of the world population and accelerated urbanization in the 21st century [16]. Smart mobility is a new and revolutionary way for a cleaner, safer and more efficient mobility, which is the most difficult issue for a smart city [8]. Since the mobility of people can be boiled down to only two categories of transport, individually owned cars and public transport (PT) systems, understanding passenger journey characteristics on a PT network is crucial not only for passenger behavior analyses, demand characterization and forecast and transit evaluation and planning [25] but also for city evaluation and planning [61, 62]. A passenger trajectory is a core element of a journey. A passenger trajectory is defined as a chronological series of location points at discrete time intervals associated with diverse context-varied attributes and characterized by a finite set of triples [57]:

$$\text{Trajectory}: \{(P_{T_1}, T_1, A_{T_1})), (P_{T_2}, T_2, A_{T_2}), \ldots, (P_{T_n}, T_n, A_{T_n}), \qquad (5.1)$$

where P_{T_n} denotes the spatial position at time T_n and A_{T_n} denotes the associated attributes, such as travel speed, heading or status.

Traditionally, traveler trajectory data were obtained commonly from surveys, such as transport surveys, mobility surveys, household surveys, population census, manual or automatic passenger count and public transportation questionnaire surveys, which were expensive to collect as well as to process [7, 11, 15, 32, 35]. However, passenger trajectory generation on a PT network was still a difficult task, as there was lack of complete and big-sample journey trajectory data for passengers. Moreover, passenger trajectory generation must capture not only the complexity of trip chaining and that of route choice on the network but also the heterogeneities during walking or waiting in a station related to individual characteristics (e.g., age, gender, burden) and to the topological complexity of each station. Novel data and approaches are requested.

In the era of Big Data, over the last decade, more diverse, accurate, abundant and large-scale new data together with related novel data processing and mining methodologies have become available owing to the deployment of Intelligent Transportation Systems (ITSs) and Information and Communications Technology (ICT) and due to the increasing computational capacities. The new data, which were collected by automatic observation systems, are named modern data. The most used modern data of a passenger on a PT network included smart ticketing data, GPS/AGPS data and mobile device data. A new branch of knowledge emerged: the modern data measurement and data-driven methods for passenger trajectory generation in the PT field [1, 30, 37, 57, 64, 65]. Indeed, traditional data modeling approaches were challenged. There is a need to summarize the knowledge of this new branch and find research trends. Only Yue *et al.* [57] proposed a state-of-the-art review of social science analyses regarding person or car trajectory-based travel behaviors rather than data-driven models for passenger trajectory generation on PT networks, in which a part of modern datasets that comply with the norm of trajectory in Equation (5.1) were discussed.

The objective of this chapter is to build a synopsis on the state-of-the-art modern datasets and data-driven models related to passenger trajectory generation methods on a PT network. The review of approaches focuses mainly on the following two aspects: (1) trip-chain generation and route choice on a PT network and (2) passenger flow characteristics in a station. A general framework combining models of the two aspects is then provided.

The benefits of the availability of passenger trajectory data for the analyses of human mobility patterns are highlighted.

This chapter is structured as follows. Section 2 gives an overview of the related modern data of passenger trajectory on a PT network. Anonymous original information provided by raw data and their limitations are summarized. In Section 3, data-driven models about passenger trajectory generation are reviewed both on a network and in a station. A general framework of data-driven models and the benefits of passenger trajectory data are depicted. Lastly, Section 4 assesses the conducted review and points to directions for further researches.

2 Overview of Related Modern Data

An overview of related modern data of passenger trajectory on a PT network collected by automatic observation systems is introduced in this section. Since a passenger journey involves passenger and vehicle movements, the data collection is divided into passenger and vehicle information measurements. For each data type, the raw data information, record device and limitations for rebuilding passenger trajectory are depicted.

2.1 *Passenger data*

Currently, modern datasets that could construct passenger trajectories on a PT network include mainly smart ticketing data and smart mobile device data.

2.1.1 *Smart ticketing data*

Smart ticketing systems were originally deployed for automated fare collection (AFC) in a transit system [37]. A smart ticket was usually electronically stored in a travel ticket on a microchip embedded on a contact or contactless smartcard and has been widely implemented all over the world since 2000. Recently, the smartphone has become both the ticketing machine and the ticket [23]. Smart ticketing data were collected by the AFC system, also called AFC data. AFC data generally included all or part of the following information: card ID (normally, anonymous ID), transaction data (status of the transaction, fare type, and fare), trip data (tap time, tap date, station ID or name, tap-gate ID, vehicle ID, run ID, transit mode, route direction and route ID), passenger personal identification data, driver information and internal database ID [1, 3, 37, 57]. Since AFC data have recently become more prevalent as a rich, comprehensive and continuous source of information on passenger journeys rather than payments, there were interesting benefits in providing individual passenger trajectories [4].

2.1.2 *Smart mobile device data*

Mobile device (mobile phone, smartphone, tablet, etc.) data collection systems were divided into two categories [12, 26]: the location system and the motion system. In the location system, radio frequency (RF) signals and satellites were used to locate the equipped mobile devices. The location system consisted of a Wi-Fi system, a bluetooth system, a GSM system and GPS/AGPS, and it recorded location information. In the motion system, mobile devices were equipped with built-in sensors, such as accelerometers, magnetometers and compasses. These recorded motion information. However, only GSM data and GPS/AGPS data and built-in sensor-based data could be used to build a passenger trajectory [54], since Wi-Fi and bluetooth data provided only fragmented information in certain places. Thus, GSM data, GPS/AGPS data and built-in sensor-based data are reviewed.

(a) GSM data

The global system for mobile communications (GSM) was mainly proposed by the telecom companies in 1990 so that subscribers could connect to cellular networks with their mobile devices and communication could occur [54]. In the meanwhile, the devices could be periodically and aperiodically localized when events such as mobile communication and internet usage took place. GSM data comprised all or part of the following information: cell ID, connected base stations, time, date, device ID and personal identification data (normally, anonymous ID) [1, 53]. The vast majority of previous experiments of GSM data considered road networks [10]. Most recently, Aguiléra *et al.* [1] were the first to use mobile phone data to study passenger individual trajectories on a six-station center trunk of the rail transit line RER A in Paris. However, the dataset was still small because of the difficulty of the transport mode in identification and the lack of GSM signal in underground stations.

(b) GPS/AGPS data

The global positioning system (GPS) using satellites was predominantly designed for global navigation by the U.S. Department of Defense 35 years ago [20]. In the middle of the 1990s, GPS data were used to save vehicle trip data by GPS devices installed in the vehicle and powered by the battery in the vehicle [50, 66]. However, only the driving trajectories were recorded before 2000. Starting in the early 2000s, traveler trip data were saved by smaller and lighter GPS devices with a detached battery, called wearable GPS devices, e.g., GeoLogger used in the U.K. in 2002 and NEVE StepLogger and Starnav used in Australia in 2003 and 2005 [18]. These wearable GPS devices still were not able to record all trips, as travelers sometimes forgot to carry the devices and GPS signal was sometimes unavailable in the building, underground (such as in underground metro stations), etc. In the middle of the 2000s, GPS function was incorporated into personal daily portable mobile devices (smartphones, smart tablets, smart watches, smart bracelets, etc.) that improved traveler trip data collection [13, 33, 38]. Otherwise, to improve the no-GPS signal environment, the assisted GPS system (AGPS) was used to connect mobile devices to satisfied GPS signals inside buildings, underground, etc. The GPS data generally included all or part of the following accurate journey information: location (latitude and longitude), time, speed, heading, the measures of data quality and personal/device information (normally, anonymous information) [43, 44]. Some methods for deriving travelers' trips from the GPS data were summarized in [18, 42]. The application of GPS data to a person or vehicle trajectory still required more accurate data.

(c) Built-in sensor-based data

Most recently, with the development of GSM and smart mobile devices (smartphone, smart tablet, smart watch, smart bracelet, etc.), built-in sensors, such as accelerometers, magnetometers and compasses, were equipped to record users' motions. Built-in sensors greatly facilitated the collection of a user's location and motion information [26]. Built-in sensor data were generally collected either through public applications that were developed by third parties or through dedicated applications designed by researchers for specific research purposes [36, 57]. Built-in sensor-based data contained all or part of the following information: acceleration, speed, time, date, location, user motion status (such as climbing up or walking down, standing, running, cycling or driving) and personal identification data (normally, anonymous ID) [14, 24, 26]. To study a traveler trajectory, sensor fusion techniques and data fusion were widely applied. Unfortunately, the sample size was still small, and there was almost no investigation along transit lines, excluding the study in a single station [53].

2.2   *Vehicle data: GPS-based AVL data*

The transit vehicle automatic measurement system is called the automatic vehicle location (AVL) system. The AVL system is a GPS-based system designed as a computer-based system for vehicle fleet control in transit systems by tracking vehicle locations in real time and has been around since the 1960s [9, 39, 40]. AFC systems shifted from ground-based radio (e.g., Loran C), signpost and dead reckoning to the present GPS-based systems with enhanced real-time location tracking and schedule monitoring. The AVL system provides all or part of the following information: vehicle location (latitude and longitude), vehicle run ID, vehicle service, line ID, arrival and departure times, date and station name [40, 55]. Thus, the vehicle trajectory based on AVL data was adjusted for inaccuracy when compared to the planned itinerary.

2.3   *Synthesis and limitations*

To sum up, those data could provide anonymous information related to passenger trajectories on a PT network. In the meantime, in handling those data, one needs to keep in mind privacy concerns, legal issues and incomplete international standardizations.

Each data has limitations for passenger trajectory generation on a PT network (Table 1). Smart mobile device data directly provide passenger trajectory points when the signals are available. However, the available data samples are still small and fragmented on the PT network. If this problem is addressed, smart mobile devices would be the main passenger trip data collection devices at a lower cost and with minimum burden. AFC data and AVL data record in-station passenger trip data and vehicle trajectory data, respectively. But they cannot directly provide personal trajectory information. Indeed, the data processing methods for modern datasets need further development.

The most recent studies have shown that the combination of AFC data and AVL data is the most feasible approach for passenger trajectory generation and application on a PT network [4] or in the origin and destination stations along a line [30, 64].

3   Passenger Trajectory Generation Approaches

Over the last decade, a new branch of knowledge has emerged: the modern data measurement and data-driven modeling for passenger trajectory generation in the PT field. AFC data and AVL data have become the

Table 1. Modern datasets and their limitations.

Modern data	Provided anonymous information[a]	Limitations
Passenger data		
AFC data	Card ID, transaction data (status of the transaction, fare type, and fare), trip data (tap time, tap date, station ID or name, tap-gate ID, vehicle ID, run ID, transit mode, route direction, and route ID), personal identification data and driver information and internal database ID	Missed tap-out stamp, transfer record, route choice, trip parts out of transit system and trip purpose
GSM data	Cell ID, location (latitude and longitude), connected base stations, time, date, device ID, and personal identification data	Missed GSM data in underground places, transportation mode and trip purpose; small sample size
GPS/AGPS data	Location (latitude and longitude), time, speed, heading and the measures of data quality, personal/device information	Missed start and end of trip, transportation mode and trip purpose; inaccuracy
Built-in sensor-based data	Acceleration, speed, time, date, location, user motion status (such as climbing up or walking down, standing, walking, running, cycling, or driving) and personal identification data	Missed data in signal weak places and trip purpose; small sample size
Vehicle data		
GPS-based AVL data	Location (latitude and longitude), time, speed, heading and the measures of data quality	Inaccuracy but adjusted according to the planned itinerary

Note: [a]Since international standardizations of the higher layers of some data categories are not widely accepted, all or part of them are provided or missed.

most used data and have been reviewed in this section. Sometimes, a PT timetable was used in place of AVL data during the first study.

Passenger trajectory generation focused mainly on on-network modeling for all transit modes. Then, it was extended recently to in-station modeling, especially the complicated stations, such as a rail transit station or a transfer hub. In general, on-network modeling (Fig. 1(a)) consists of two phases: trip-chain generation and route choice including vehicle run choice (departure time choice), transfer choice, itinerary choice and mode choice. In-station modeling (Fig. 1(b)) is about passenger flow characteristics. Thus, passenger trajectory in the transit system consists of different stages:

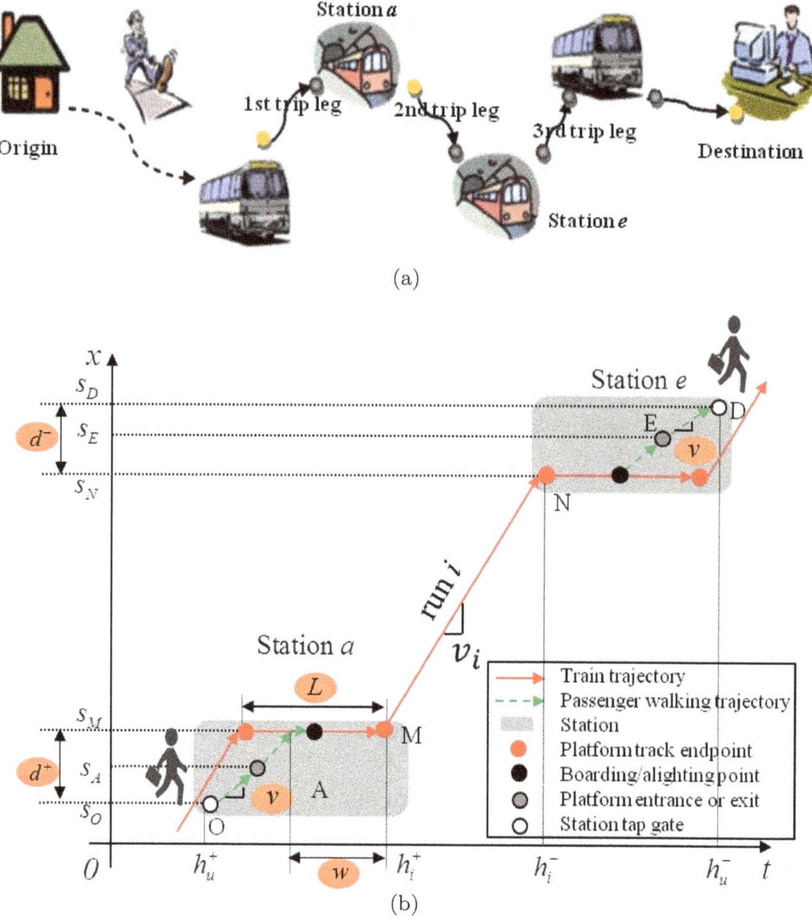

Fig. 1.   Passenger movements: (a) on-network and (b) in stations on the second trip leg.

passenger arrival/entry (tap-in), walking-in, wait, in-vehicle, walking-out, passenger exit (tap-out), with or without transfer.

3.1   *On-network trip-chain and route choice*

The first key element of on-network transit modeling is passenger journey generation composed of several origin–destination (OD) trips (Fig. 1(a)), also called trip legs. To accomplish this, trip-chaining methods constructed a passenger's travel sequence and mode choice with specific assumptions (studied in [3, 5, 19, 31, 34, 52] and first validated in [4]). Since most AFC systems are tap-in only systems, meaning no tap-out records while alighting and no trip destination information, destination inference algorithms were needed. In the meantime, the trip transfer information was inferred. The generated passengers' individual trajectories were further used for passenger trip purpose inference when combined with other datasets [4]. However, studies of passenger exit in egress station and investigations of multimodal transport including other modes rather than PT were few. Also, trip-chaining methods focused mainly on inferences of trip destination, and transfer elements, train run matching and route choice strategies with vehicle capacity constraint were not explicitly considered.

The second key element of on-network transit modeling is the probability of matching between a passenger journey and a train run serving the boarding station. On a trip leg, a methodology to infer rail transit OD matrix and generate path choice from Oyster smartcard data in London was proposed in [60]. The authors used data processing and analysis methods supported by technologies of database management systems (DBMS) and geographic information systems (GIS). Based on this method and a Poisson arrival assumption, a mixed-weight inference of passenger path choice behavior, gate-to-gate journey time distribution and crowding metrics on a platform were derived from the AFC data between two OD pairs in Beijing. Based on this study, Sun and Schonfeld [47] further developed a schedule-based assignment model by extending the fail-to-board (or left behind) model of [41] and studying the transfer penalty. Since simplified assumptions were made, these constrained the model application. To mitigate this drawback, Zhu *et al.* [64] devised a probabilistic individual Passenger-to-Train Assignment Model (PTAM), in which the matching probability between train runs and individual journeys was emphasized. The model was designed for the fusion of AFC data and AVL data on a five-station segment of metro line in Hong Kong and validated by synthetic data. The PTAM was developed so as to provide estimators of the main state variables of passenger flows both in station (waiting loads) and in vehicle (loading snakes) and that of several factors of transit service, especially the in-station crowding level. It was further applied to the estimation of the distribution of fail-to-board (left behind) on the boarding platform (the number of trains/times a passenger was unable to board due to capacity constraint) [65]. Nevertheless, this model considered only a two-itinerary choice scenario without transfer.

Besides train run choice, passengers' route choice modeling with transfer on a complex large-scale transit network was a difficult task as well. Zhao *et al.* [59] introduced a comprehensive literature review of previous route and train run choice approaches: previous approaches of passengers' route and train run choice modeled only some specific situations. This means the approaches did not consider complicated situations: (i) heavy "left-behinds" in a series of stations and (ii) long walking times in access and transfer stations. The authors developed a probabilistic model that simulated how passenger flows match to different routes and trains from empirical analysis combining AFC data and AVL data on a complex large-scale metro network in Shenzhen. In the meantime, to integrate transfer in crowding discomfort analysis, Hörcher *et al.* [21] combined the PTAM model of Zhu *et al.* [64] with the standing and sitting choice models of Leurent [27] and Sumalee *et al.* [45]. This model was applied to a metro network in Hong Kong. A group of transfer choices was modeled in a Bayesian framework at individual journey level for complex trips. This involved transfers, hence combined legs (of journeys), in order to study the route choice between station pairs and revealed the influence of in-vehicle congestion on route disutility to passengers. However, in-station passenger flow features were scarcely considered in on-network models.

3.2 *In-station passenger flow*

At boarding stations, the individual passenger arrival was investigated by trip leg in [17, 58]. Two types of passenger arrivals, timetable-dependent arrival and random arrival, were distinguished and subsequently used to evaluate the actual realized and excess journey times. However, the authors did not explicitly model the statistical properties of the passenger arrival distributions. Ingvardson *et al.* [22] proposed an approach explicitly taking into account the timetable-dependent arrival and random arrival as uniform beta distributions to model passenger waiting time as a mixture distribution. Within this scope, the issue of passenger positioning on the boarding station was mentioned in [63]. It was suggested that this be modeled as a random variable, yet apparently no further development took place. Passenger longitudinal positioning distance was primarily modeled in [29] and extended with posterior analysis and train run choice in [30]. To simplify computational complexity and save model optimization time, the passenger longitudinal positioning distance was modeled explicitly, but waiting times was modeled implicitly.

In OD stations on a trip leg, stochastic features of passenger journeys between tap-in and tap-out moments were modeled in separate stages: walking-in, wait, in vehicle, walking-out, with or without transfer between OD pairs [48]. A regression model was proposed to generate distributions of walking-in and walking-out times in access and egress stations (gamma distributions) and distributions of waiting times at access and transfer stations (uniform distributions) including the additional waiting time caused by fail-to-board. This approach was further applied in the PTAM in [64]. Nevertheless, this study considered explicitly only distributions of walking-in and walking-out speeds (normal distributions). To refine this study, econometric methods were also applied for modeling and estimating individual passenger in-station underlying variables. Uniform-distributed passenger walking speed and shifted exponential-distributed distances by trip leg along passenger trajectories were modeled explicitly in passenger flow stochastic models (PFSMs) by Leurent and Xie [28]. Parameters were estimated by the maximum likelihood estimation (MLE) combining AFC data and the timetable of the line RER A in the Paris area. Xie and Leurent [55] demonstrated that normal-distributed walking speed reduced a lot of the complexity of the likelihood function computation. The model was further developed by adding the waiting time component as a shifted exponential distribution in [56]. The extended model was computation tractable and given consistent estimates with the former. The stochastic models and related MLE were more analogous to those in [64]. Nevertheless, the consideration of passenger individual distance in the latter resolved the conservative assumption that the minimum access and egress time were zero (infinite passenger speed) in [46, 47, 64, 65]. Simultaneously, Wahaballa *et al.* [51] provided a stochastic frontier model extending the model in [2] to estimate Erlang-distributed waiting time and service reliability. Walking times and in-vehicle travel times were normal-distributed distributions. Parameters were estimated by the MLE method using AFC data and the timetable of the Piccadilly line of London's underground network.

3.3 *Synthesis and benefits*

Based on the aforementioned methods, a general framework of data-driven models for passenger trajectory generation and application is summarized in Table 2. The potential applications and side products are suggested.

The availability of passenger trajectory data improves the understanding of the transit and city functions. For the PT, the passenger behavior analysis can be extended to in-station underlying indicators which were difficult to measure [21, 30]. For the city, we can investigate the human mobility patterns to understand urban dynamics on a large scale, where there was lack of data before [62]. Otherwise, during transit demand characterization and forecast, the dynamic OD matrix of a PT of a city can be generated from the passenger trajectories [3]. More accurate mobility activity study of a big sample including data on travel purpose

Table 2. General framework for passenger trajectory generation and application.

Input:	Individual passenger data and/or vehicle data
Step 1:	Data clearance and preprocessing: passenger data, vehicle data and transit network information
Step 2:	Passenger trajectory generation combining passenger data, vehicle data and transit network information • On a network: trip-chain and route choice (including vehicle run choice, transfer choice, itinerary choice and mode choice) • In a station: passenger flow characteristics
Step 3:	Further applications: passenger behavior analyses; demand characterization and forecast; transit and city evaluation and planning
Output:	(1) Passenger flow, vehicle flow, transit network functions, city network functions; (2) decision-making indicators
Side products:	Modern data exploitation and management systems and decision-making support tools

can be achieved [4]. The demand of a PT can be forecasted from real-time streaming [49]. Finally, passenger flow, vehicle flow, transit network functions, city network functions and related decision-making indicators will be the main bricks to develop smart mobility technologies such as smart traffic management, urban data platform, sharing services layer (apps), smart transit (bus on demand), transit infrastructure expenditure forecast, etc. [16] in the context of the smart city.

4   Conclusions and Future Research Directions

After reviewing the related datasets and recent data-driven models for passenger trajectory generation on a PT network, in this section, a synthesis is given, and some research directions are pointed out.

4.1   *Conclusion*

Over the last decade, new and massive data emerging in the PT system brought both challenges and opportunities for research and application. A new branch of knowledge, modern data, measurement and data-driven modeling in the PT field, emerged. A major breakthrough was made in passenger trajectory generation using data-driven models. This chapter reviewed the related modern data and recent data-driven models for passenger trajectory generation on a PT network.

An overview of modern data related to passenger trajectory on the network was proposed. In handling those data, one should keep in mind privacy concerns, legal issues and incomplete international standardizations. In most studies, anonymous information was provided. The limitations of all reviewed data were summarized in Table 1. To summarize, AFC data and AVL data (or timetables) became the most used datasets in current studies, and mobile device data would be the next most used data while their samples become bigger and more complete.

Data-driven modeling consisted of on-network and in-station modeling. Data-driven models tried to capture the congestion phenomena: either in-vehicle crowding (the potential restrictions), i.e., the probability of fail-to-board (the related issue of passengers "left behind" by trains), or the crowding on platforms and its potential influence on individual positioning, but not the crowding at platform access and egress points, which may entail queuing and delay among exiting passengers.

A general framework of a data-driven model for passenger trajectory generation is shown in Table 2. Potential applications and side products were also suggested. The availability of passenger trajectory data brought benefits for the following: (1) the passenger behavior analysis, (2) demand characterization and forecast, (3) transit evaluation and planning and (4) city evaluation and planning.

4.2 *Future research directions*

Four future directions can be identified in relation to gaps in passenger trajectory generation and application.

Since the investigation of multimodal transport on a network was limited, the development of data-driven multimodal transport modeling (including PT, private vehicle, walking, etc.) makes up the first direction.

A second direction is to devote more attention to the in-station phases, access and egress stations, especially so for vertical pedestrian elements that influence individual speed under congestion as well as free-flow conditions. These issues are well known in the micro-simulation of pedestrian traffic (e.g., simulation software, such as the Legion or Viswalk), but their estimations on the basis of modern data are still an open issue.

Third, more detailed data of passengers' trajectories will be available from smart mobile devices owing to applications that monitor geolocation data and motion data from one or several sources: GPS, GSM or built-in sensors. Indeed, location data and motion data collected every second constitute ideal material for the refined analysis of passenger trip-making. Such research remains to be done for large underground transit stations, where satellite or beacon signals are impeded or modified by local layout corridors, walls, floor or ceilings. Indeed, the data processing methods for modern datasets need further development to accomodate new and multiple data sources.

Lastly, more accurate passenger trajectories will aid the development of new smart mobility technologies in the context of the smart city. This will be the backbone for smart mobility development.

References

[1] Aguiléra, V., Allio, S., Benezech, V., Combes, F., & Milion, C. (2014). Using cell phone data to measure quality of service and passenger flows of Paris transit system. *Transportation Research Part C: Emerging Technologies*, 43, 198–211.

[2] Aigner, D., Lovell, C., & Schmidt, P. (1977). Formulation and estimation of stochastic frontier production function models. *Journal of Econometrics*, 6(1), 21–37. https://doi.org/10.1016/0304-4076(77)90052-5.

[3] Alsger, A., Assemi, B., Mesbah, M., & Ferreira, L. (2016). Validating and improving public transport origin-destination estimation algorithm using smart card fare data. *Transportation Research Part C: Emerging Technologies*, 68, 490–506. doi:10.1016/j.trc.2016.05.004.

[4] Alsger, A., Tavassoli, A., Mesbah, M., Ferreira, L., & Hickman, M. (2018). Public transport trip purpose inference using smart card fare data. *Transportation Research Part C: Emerging Technologies*, 87, 123–137. doi: 10.1016/j.trc.2017.12.016.

[5] Alsger, A. A., Mesbah, M., Ferreira, L., & Safi, H. (2015). Use of smart card fare data to estimate public transport origin–destination matrix. *Transportation Research Record: Journal of the Transportation Research Board*, 2535, 88–96. doi: 10.3141/2535-10.

[6] Alsger, A. A. M. (2016). Estimation of transit origin destination matrices using smart card fare data. Ph.D. Thesis, University of Queensland, 1–206.

[7] Beatty, C., & Haywood, R. (1997). Changes in travel behaviour in the English Passenger Transport Executives' areas 1981–1991. *Journal of Transport Geography*, 5(1), 61–72.

[8] Benevolo, C., Dameri, R. P., & D'Auria, B. (2016). Smart mobility in smart city action taxonomy, ICT intensity and public benefits. *Lecture Notes in Information Systems and Organisation*, 11, 13–28. doi: 10.1007/978-3-319-23784-8_2.

[9] Cain, D. A., & Pekilis, B. R. (1993). AVLC technology today: A developmental history of automatic vehicle location and control systems for the transit environment. In *Proceedings of VNIS '93 — Vehicle Navigation and Information Systems Conference*. doi: 10.1109/VNIS.1993.585700.

[10] Chen, C., Ma, J., Susilo, Y., Liu, Y., & Wang, M. (2016). The promises of big data and small data for travel behavior (aka human mobility) analysis. *Transportation Research Part C: Emerging Technologies*, 68, 285–299.

[11] Chen, X., Liu, Q., & Du, G. (2011). Estimation of travel time values for urban public transport passengers based on SP survey. *Journal of*

Transportation Systems Engineering and Information Technology, 11(4), 77–84.

[12] Choujaa, D., & Dulay, N. (2009). Activity recognition from mobile phone data: State of the art, prospects and open problems. Imperial College London, V, 1–32.

[13] City, M., Shinji, I., Division, T. E., & Management, I. (2006). Development of MoALs (Mobile Activity Loggers supported by gps-phones) for travel behavior analysis Eiji HATO. *Transportation Research Record*, August 2005.

[14] Deutsch, K., & Mckenzie, G. (2012). Examining the use of smartphones for travel behavior data collection. In *13th International Conference on Travel Behaviour Research*, July, Toronto.

[15] Du, P., Liu, C., & Liu, Z. (2009). Walking time modeling on transfer pedestrians in subway passages. *Journal of Transportation Systems Engineering and Information Technology*, 9(4), 103–109.

[16] EIP-SCC (2016). Towards a joint investment programme for European smart cities: A consultation paper to stimulate action, 1–28.

[17] Frumin, M., & Zhao, J. (2012). Analyzing passenger incidence behavior in heterogeneous transit services using smartcard data and schedule-based assignment. *Transportation Research Record: Journal of the Transportation Research Board*, 1, 52–60.

[18] Gong, L., Morikawa, T., Yamamoto, T., & Sato, H. (2014). Deriving personal trip data from GPS data: A literature review on the existing methodologies. *Procedia — Social and Behavioral Sciences*, 138, 557–565. doi: 10.1016/j.sbspro.2014.07.239.

[19] Gordon, J., Koutsopoulos, H., Wilson, N., & Attanucci, J. (2013). Automated inference of linked transit journeys in London using fare-transaction and vehicle location data. *Transportation Research Record: Journal of the Transportation Research Board*, 2343, 17–24. doi: 10.3141/2343-03.

[20] He, X., Montillet, J. P., Fernandes, R., Bos, M., Yu, K., Hua, X., & Jiang, W. (2017). Review of current GPS methodologies for producing accurate time series and their error sources. *Journal of Geodynamics*, 106, 12–29. doi: 10.1016/j.jog.2017.01.004.

[21] Hörcher, D., Graham, D. J., & Anderson, R. J. (2017). Crowding cost estimation with large scale smart card and vehicle location data. *Transportation Research Part B: Methodological*, 95(Supplement C), 105–125.

[22] Ingvardson, J. B., Nielsen, O. A., Raveau, S. and Nielsen, B. F. (2018). Passenger arrival and waiting time distributions depending on train service frequency and station characteristics: A smart card data analysis, *Transportation Research Part C: Emerging Technologies*, 90, 292–306.

[23] ITSO (2018). Integrated transport smartcard organisation, who maintain the ITSO standard for smart ticketing in the UK https://www.itso.org.uk/.

[24] Kim, J. S., Gračanin, D., & Quek, F. (2012). Sensor-fusion walking-in-place interaction technique using mobile devices. *Proceedings — IEEE Virtual Reality*, 39–42. doi: 10.1109/VR.2012.6180876.

[25] Kurauchi, F., & Schmöcker, J.-D. (2016). *Public Transport Planning with Smart Card Data*. CRC Press, 1–261.

[26] Lane, N. D., Miluzzo, E., Lu, H., Peebles, D., & Choudhury, T. (2010). A survey of mobile phone sensing. *IEEE Communication Magazine*, 140–150.

[27] Leurent, F. (2008). Modelling seat congestion in transit assignment, Working document made available at HAL, 1–28.

[28] Leurent, F., & Xie, X. (2017). Exploiting smartcard data to estimate the distributions of passengers' walking speed and distances along an urban rail transit line. *Transportation Research Procedia*, 22, 45–54.

[29] Leurent, F., & Xie, X. (2017). On passenger repositioning along station platform during train waiting. *Transportation Research Procedia*, 27, 688–695. doi: 10.1016/j.trpro.2017.12.123.

[30] Leurent, F., & Xie, X. (2018). On individual repositioning distance along platform during train waiting. *Journal of Advanced Transportation*, 1–18. doi: https://doi.org/10.1155/2018/4264528.

[31] Li, T., Sun, D., Jing, P., & Yang, K. (2018). Smart card data mining of public transport destination: A literature review. *Information*, 9(1), 18. doi: 10.3390/info9010018.

[32] Long, Y., & Thill, J.-C. (2015). Combining smart card data and household travel survey to analyze jobs-housing relationships in Beijing. *Computers, Environment and Urban Systems*, 1–35.

[33] Millward, H., Spinney, J., & Scott, D. (2013). Active-transport walking behavior: Destinations, durations, distances. *Journal of Transport Geography*, 28, 101–110. doi: 10.1016/j.jtrangeo.2012.11.012.

[34] Nassir, N., Khani, A., Lee, S., Noh, H., & Hickman, M. (2011). Transit stop-level origin-destination estimation through use of transit schedule and automated data collection system. *Transportation Research Record: Journal of the Transportation Research Board*, 2263, 140–150. doi: 10.3141/2263-16.

[35] Nguyen, S., Morello, E., & Pallottino, S. (1988). Discrete time dynamic estimation model for passenger origin/destination matrices on transit networks. *Transportation Research Part B: Methodological*, 22(4), 251–260.

[36] Nitsche, P., Widhalm, P., Breuss, S., Brändle, N., & Maurer, P. (2014). Supporting large-scale travel surveys with smartphones — A practical approach. *Transportation Research Part C: Emerging Technologies*, 43, 212–221. doi: 10.1016/j.trc.2013.11.005.

[37] Pelletier, M.-P., Trépanier, M., & Morency, C. (2011). Smart card data use in public transit: A literature review. *Transportation Research Part C: Emerging Technologies*, 19(4), 557–568. doi: 10.1016/j.trc.2010.12.003.

[38] Pereira, F., Carrion, C., Zhao, F., Cottrill, C. D., Zegras, C., & Ben-Akiva, M. E. (2013). The Future Mobility Survey: Overview and preliminary evaluation. *Proceedings of the Eastern Asia Society for Transportation Studies*, 9, 1–13.

[39] Riter, S., & McCoy, J. (1977). Automatic vehicle location — An overview *IEEE Transactions on Vehicular Technology*, 26(1), 7–11.

[40] Sandim, M., Rossetti, R. J. F., Moura, D. C., Kokkinogenis, Z., & Rúbio, T. R. P. M. (2016). Using GPS-based AVL data to calculate and predict traffic network performance metrics: A systematic review. In *Proceedings of the IEEE Conference on Intelligent Transportation Systems, Proceedings, ITSC*, 1692–1699. doi: 10.1109/ITSC.2016.7795786.

[41] Schmöcker, J. D. (2006). Dynamic capacity constrained transit assignment. Ph.D. Thesis, Imperial College London, London, p. 202. http://www3. imperial.ac.uk/pls/portallive/docs/1/36049696.PDF.

[42] Shen, L., & Stopher, P. R. (2014). Review of GPS travel survey and GPS data-processing methods. *Transport Reviews*, 34(3), 316–334. doi: 10.1080/01441647.2014.903530.

[43] Stopher, P., FitzGerald, C., & Zhang, J. (2008). Deducing mode and purpose from GPS data. *87th Annual Meeting of the Transportation Research Board*, 1–11.

[44] Stopher, P., FitzGerald, C., & Zhang, J. (2008). Search for a global positioning system device to measure person travel. *Transportation Research Part C: Emerging Technologies*, 16(3), 350–369. doi: 10.1016/j.trc.2007.10.002.

[45] Sumalee, A., Tan, Z., & Lam, W. H. K. (2009). Dynamic stochastic transit assignment with explicit seat allocation model. *Transportation Research Part B: Methodological*, 43(8–9), 895–912. doi: 10.1016/j.trb.2009.02.009.

[46] Sun, L., Lee, D.-H., Erath, A., & Huang, X. (2012). Using smart card data to extract passenger's spatio-temporal density and train's trajectory of mrt system. In *Proceedings of the ACM SIGKDD International Workshop on Urban Computing*, 142–148.

[47] Sun, Y., & Schonfeld, P. M. (2016). Schedule-based rail transit path-choice estimation using automatic fare collection data. *Journal of Transportation Engineering*, 142(1), 04015037. doi: 10.1061/(ASCE)TE.1943-5436.0000812.

[48] Sun, Y., & Xu, R. (2012). Rail transit travel time reliability and estimation of passenger route choice behavior. *Transportation Research Record: Journal of the Transportation Research Board*, 2275, 58–67. doi: 10.3141/2275-07.

[49] Toque, F., Come, E., Oukhellou, L., Trepanier, M., Toqué, F., Côme, E., & Trépanier, M. (2018). Short-term multi-step ahead forecasting of railway passenger flows during special events with machine learning methods. In *CASPT 2018, Conference on Advanced Systems in Public Transport and TransitData 2018*, July 2018, Brisbane, Australia, 15–2018. https://hal. archives-ouvertes.fr/hal-01891446.

[50] Wagner, D. P. (1997). Lexington area travel data collection test: GPS for personal travel surveys. Final Report, Office of Highway Policy Information and Office of Technology Applications, Federal Highway Administration, Battelle Transport Division, Columbus, pp. 1–92.

[51] Wahaballa, A. M., Kurauchi, F., Yamamoto, T., & Schmöcker, J.-D. (2017). Estimation of platform waiting time distribution considering service reliability based on smart card data and performance reports. *Transportation Research Record: Journal of the Transportation Research Board*, 2652, 30–38.

[52] Wang, W. (2010). Bus passenger origin-destination estimation and travel behavior using automated data collection systems in London, UK. *Master Thesis of Massachusetts Institute of Technology*, 1–130.

[53] Wang, Z., He, S. Y., & Leung, Y. (2018). Applying mobile phone data to travel behaviour research: A literature review. *Travel Behaviour and Society*, 11, 141–155. doi: 10.1016/j.tbs.2017.02.005.

[54] Wolf, J., Bachman, W., Oliveira, M., Auld, J., Mohammadian, A., Vovsha, P., & Zmud, J. (2014). *Applying GPS Data to Understand Travel Behavior*, Vol. 1. Transportation Research Board, p. 529.

[55] Xie, X., & Leurent, F. (2017). Comparison of passenger walking speed distribution models in mass transit stations. *Transportation Research Procedia*, 27, 696–703.

[56] Xie, X., & Leurent, F. (2017). Estimating distributions of walking speed, walking distance, and waiting time with automated fare collection data for rail transit. *Transportation Research Record: Journal of the Transportation Research Board*, 2648(2), 134–141.

[57] Yue, Y., Lan, T., Yeh, A. G. O., & Li, Q. (2014). Zooming into individuals to understand the collective: A review of trajectory-based travel behaviour studies. *Travel Behaviour and Society*, 1(2), 69–78.

[58] Zhao, J., Frumin, M., Wilson, N., & Zhao, Z. (2013). Unified estimator for excess journey time under heterogeneous passenger incidence behavior using smartcard data. *Transportation Research Part C: Emerging Technologies*, 34(1), 70–88.

[59] Zhao, J., Zhang, F., Tu, L., Xu, C., Shen, D., Tian, C., Li, X.-Y., & Li, Z. (2017). Estimation of passenger route choice pattern using smart card data for complex metro systems. *IEEE Transactions on Intelligent Transportation Systems*, 18(4), 790–801.

[60] Zhao, J., Rahbee, A., & Wilson, N. H. M. (2007). Estimating a rail passenger trip origin-destination matrix using automatic data collection systems. *Computer-Aided Civil and Infrastructure Engineering*, 22(5), 376–387.

[61] Zhong, C., Arisona, S. M., Huang, X., Batty, M., & Schmitt, G. (2014). Detecting the dynamics of urban structure through spatial network analysis. *International Journal of Geographical Information Science*, 28(11), 2178–2199. doi: 10.1080/13658816.2014.914521.

[62] Zhong, C., Batty, M., Manley, E., Wang, J., Wang, Z., Chen, F., & Schmitt, G. (2016). Variability in regularity: Mining temporal mobility patterns in London, Singapore and Beijing using smart-card data. *PLoS ONE*, 11(2). doi: 10.1371/journal.pone.0149222.

[63] Zhu, Y. (2014). Passenger-to-train assignment model based on automated data. MSC Thesis, S.M. Massachusetts Institute of Technology, 1–113. http://dspace.mit.edu/handle/1721.1/90075.

[64] Zhu, Y., Koutsopoulos, H. N., & Wilson, N. H. M. (2017). A probabilistic passenger-to-train assignment model based on automated data. *Transportation Research Part B: Methodological*, 104, 522–542. doi: https://doi.org/10.1016/j.trb.2017.04.012.

[65] Zhu, Y., Koutsopoulos, H. N., & Wilson, N. H. M. (2017). Inferring left behind passengers in congested metro systems from automated data. *Transportation Research Procedia, Transportation Research Part C: Emerging Technologies*, 23(Supplement C), 362–379. doi: 10.1016/j.trpro.2017.05.021.

[66] Zito, R., D'Este, G., & Taylor, M. A. P. (1995). Global positioning systems in the time domain: How useful a tool for intelligent vehicle-highway systems? *Transportation Research-C*, 3(4), 193–209.

Chapter 6

Toward Autonomous Valet Parking: A Broader Perspective

Muhammad Khalid*, Yue Cao[†], Kezhi Wang*,
Piyush Dhawankar* and Mohsin Raza[‡,§]
*Department of Computer & Information Sciences,
Northumbria University, Newcastle Upon Tyne, UK
[†]School of Computing & Communication,
Lancaster University, UK
[‡]Department of Design Engineering & Mathematics,
Middlesex University, UK
[§]m.raza@mdx.ac.uk

1 Overview of the Chapter

Continuous and effective developments in autonomous vehicles (AVs) are happening on a daily basis. Industries nowadays are interested in introducing less expensive and highly controllable AVs to the public. The current so-called autonomous valet parking (AVP) solutions are still limited to a very short range (e.g., they may only work at the entrance of car parks (CPs)). This chapter proposes a parking scheduling scheme for a long-range AVP (LAVP) case by considering the mobility of the AVs, fuel consumption and journey time. In LAVP, CPs are used to accommodate increasing numbers of AVs and are placed outside the city center in order to avoid traffic congestions and ensure road safety in public places. Furthermore, with the positioning of reference points to guide user-centric long-term driving and passenger drop-off/pick-up, simulation results under the Helsinki city scenario show the benefits of LAVP. The advantage of the LAVP system is also reflected through both analysis and simulation.

2 Introduction

The transportation system is an important aspect of human life these days. Mobility is largely dependent on the transportation system irrespective of whether it is public or private. According to the Transport Statistics Great Britain 2017 [9], 78% of the distance traveled by a person is covered by personal transport while the remaining 22% is covered by other means of public transport. As the recent statistics show, there is an increase in the private means of transport. People tend to travel by their own vehicles due to better comfort. Recent studies state that vehicles are parked for 95% of their lifetime. As per the British National Travel Survey [5], a car is driven an average of only 361 hours a year. According to a recent survey, drivers spend almost 2,549 hours (almost 106 days of life) to search for parking space. In the UK, it takes an average of around 405 seconds to find a proper parking space, according to a survey issued by JustPark [12]. It is normally seen that due to parking challenges, around 32% of people prefer to take the public transport to work. As each and every parking slot has very limited space and strict dimensions, most of the time, one needs to move the car forward and backward multiple times to park accurately. Parking a car exactly within the borders of a parking slot is a challenging job.

Recent inventions in information communication technology (ICT), sensing and distributed control systems have solved many of the transportation issues regarding traffic congestion, efficiency, safety on wheels and environmental pollution. Driving experience has been enhanced through the development of various smart applications in intelligent transportation systems (ITSs).

It is observed that around 20% of accidents occur at CPs. Parking a vehicle in commercial parking needs expert driving skills. In many cities, traffic administrations have developed parking guidance and information systems (PGIs) to facilitate AVs on the road. A PGI uses sensors to detect and monitor vehicles in the parking area. These sensors are usually classified into two categories: in-roadway sensors and out-roadway sensors [19].

To enhance the transportation system with the existing infrastructure and increasing number of vehicles, intelligent mechanisms are needed. "Park and ride" has reduced the burden of parking in congested areas. Drivers who park their cars in remote CPs walk toward the city center, drivers using long-term parking will be charged with a lower cost and sometimes parking will be offered without any cost. On the contrary, an extra cost will be required to travel between remote CPs and drop-off zones. Till now, factors regarding parking have been addressed only in terms of cost; however, the precious time of drivers will be spent in driving cars to remote CPs and then using public transport. Similarly, one needs to travel back to remote CPs to drive their car toward their destination. Usually, "park and ride" is adopted to lower the congestion rate in public areas. Consider the case where the drivers park their vehicles in a remote CP and walk down or travel by bus to the work place (WP). In this scenario, in terms of cost, the driver will have either free parking or parking with a much lower fee. But, the user will pay bus fare to travel to the WP. On the other hand, extra time will be needed, such as the time required to walk toward the bus stop from the CP and the waiting time at the bus stop. However, this methodology has environmental as well as economic benefits. In return, drivers need to sacrifice some of their precious time. Recent improvements in machine learning (ML) systems and autonomous car-maneuvring techniques in short-range autonomous valet parking (SAVP) are helping drivers in routine parking, as depicted in Fig. 1. Recent advancements in machine vision and maneuvering systems can help drivers to park vehicles autonomously with accurate measures and efficiency. SAVP has also improved the quality of the driving experience. SAVP can help to park a car autonomously in simple and trained scenarios. Due to ML techniques, an AV learns to park autonomously after a trial which should be performed by the driver at least once. In advanced AVP mechanisms, AVs can scan for empty parking slots in a multistory carpark as well.

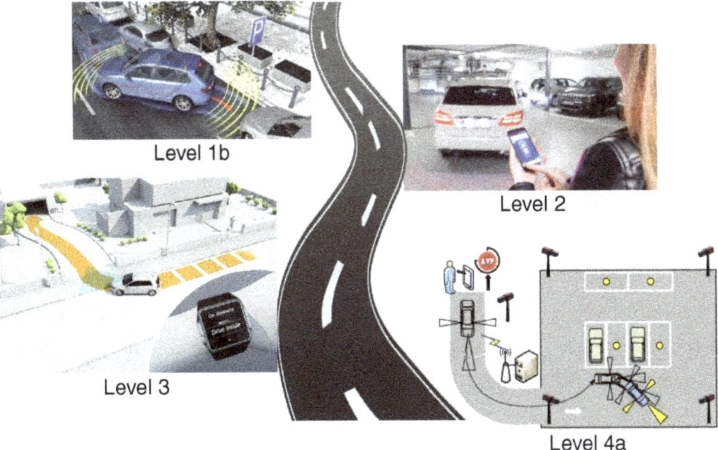

Fig. 1. Different steps involved in autonomous valet parking (AVP).

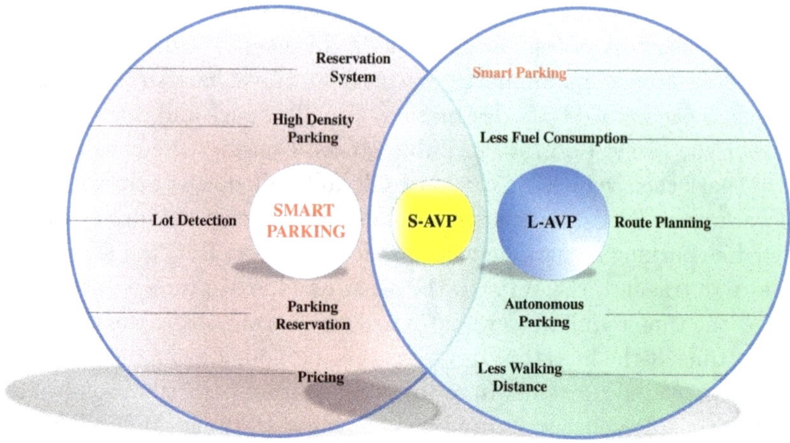

Fig. 2. Transition from smart parking to AVP.

In recent years, a lot of development has taken place in AVs. As a result of developments in ICT, CPs' status throughout the city is known to all AVs. A clear figure regarding transition from smart parking to AVP has been illustrated into Fig. 2, current information about CPs, such as parking fee, opening and closing hours, available slots and distance from the city center/drop-off zone, are also available [21]. A few of the capabilities of AV may be breakthroughs for building a smart transportation system with existing infrastructure. AVs offer more flexible and efficient solutions for transportation systems. In LAVP, a driver selects the drop-off point and leaves the vehicle over there. A CP is selected from a range of CPs through a mobile application. The selection of the CP is dependent on various parameters, like distance from drop-off spot, parking fee, traffic situation and road conditions. The vehicle moves forward autonomously with the help of a vehicle controller (VC) and can be monitored by the user through a mobile device. The AV is parked in a selected CP. Upon request, the vehicle starts moving again toward the selected pick-up spot. Similarly, the vehicle is picked up by the driver at the pick-up spot, as shown in Fig. 3.

For each of the problems explained above, various techniques have been proposed. However, the following aspects must be considered in making parking easy and effective:

- How can we overcome congestion in urban areas?
- How can we ensure safe and hassle-free parking?
- How can the cost be reduced for long-term parking?

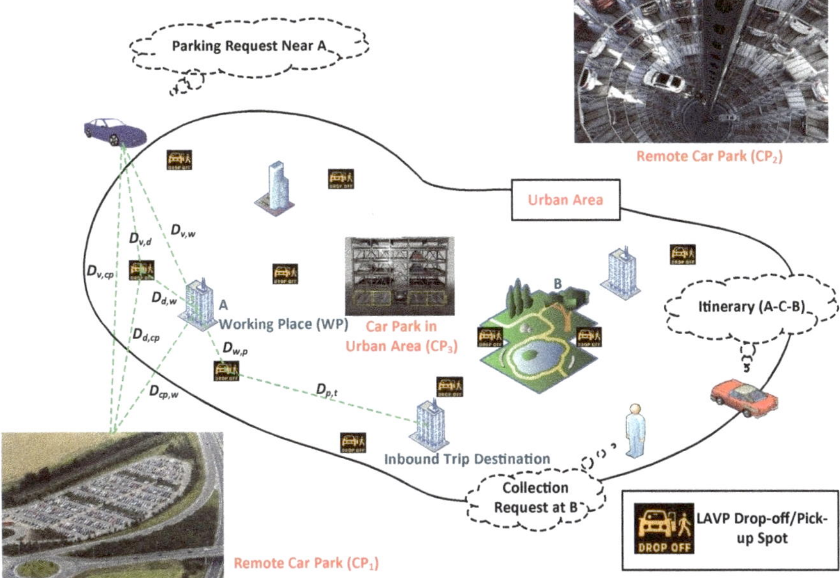

Fig. 3. The framework for LAVP.

- How efficiently can parking lots be utilized?
- How reliable will the parking be?

In this chapter, we have proposed a new cost estimation and scheduling model for LAVP. Through this model, we can form an estimate of the parking fee for a specific CP. In this system, the parking fee is mainly dependent on how far the CP is from the drop-off zone, how much time is required to reach the CP and the time slot for which parking will be taken.

3 Background on AVP

The idea of AVs was mainly developed around the 1920s, while research has been carried out for more than 30 years to bring those ideas into practical use. Usually, AVs are equipped with sensors similar to those of underwater autonomous vehicles (UAVs) and can sense the nearby environment in the range of the sensors [14, 17]. The concept of autonomy is also used in UAVs [18]. It enables the UAV to move freely without human intervention [3, 16].

Considerable work has been carried out to study the behavior of current parking scenarios and improve the parking efficiency [7, 15]. In the earlier stages of research in this specific area, a number of models have been introduced. The models developed to understand the choice of parking include PARKSIM, CLAMP, PARKAGENT and many other models [1, 25].

In [2], an automatic valet parking system using a nomadic device and parking server has been introduced. This system consists of four major parts: a nomadic device, a VC, a local spatial-aware server and a global situation-aware server. A nomadic device is usually a mobile phone or tablet that is mounted on the vehicle dashboard. This device is used to operate a parking application. VC receives maps, control commands and other traffic information from the local spatial-aware server. The local spatial-aware server detects an obstacle in the vehicle's way and sends control commands to the vehicle accordingly, while the global situation-aware server makes high-level path-planning decisions as per the driving environment. In [23], the proposed system uses AVP mobile, the AVP VC and the AVP server for parking operations. AVP mobile has a parking application while the AVP VC controls the vehicle on road. The AVP server is input with updated information about parking space and it sends control commands to the VC as well. This scheme uses geography mark-up language (GML) to model the geography of the parking spaces. In [11], a smart parking system has been introduced. A new parking system called i-Parker is introduced in [20]. It uses the concept of intelligent resource allocation, less pricing and in-time reservation (see Fig. 2).

AVP delivers astonishing services with the help of modern automation technologies. It improves the overall user experience and provides safety as well. AVP is evolving along with automation technologies and ICT. It provides services at different levels [10].

In the initial days, an AVP was used to provide limited parking assistance. Automatic parking is used while the driver remains in the AV, which is not fully autonomous as the driver can intervene in the process. In this process, the whole parking activity is fully supervised by the driver, and this is referred to as "Level 1b" in Fig. 1. In this process, the driver drives the AV to a vacant parking lot and sets its position at a certain distance from obstacles and other AVs. Once the AV is in heading position toward the parking lot, it automatically detects the lot and is parked. This mechanism is mostly useful for a less-experienced driver, and it has minimum chances of hitting an obstacle or other AVs.

In the following years, wireless operations were developed for use within the AVP system. This enables the driver to stay out of the car, and to perform and monitor the parking process through their specified handset or smart phone. This is referred to as "Level 2".

In the later stage, which is shown as "Level 3", three-dimensional mapping and sensing technologies are used. This is a more advanced level of AVP than the previous one, where AVs travel to a parking lot from a specific spot. Usually, in these techniques, an AV is trained at least once with the driver inside the AV [8].

In the state-of-the-art AVP system, path generation [22] and precise detection [6] techniques have extended AVP's scope to large-scale areas. In the scenario referred to as "Level 4a", a driver leaves their AV at the entrance of CP and navigates the AV toward a vacant lot [2]. The disadvantage of this system is that the driver has to approach the CP and drop the AV over there; however, it saves the time taken to find a lot and park autonomously in the CP.

4 Design for LAVP

The exponential increase in the number of vehicles has increased the parking difficulties in urban and congested areas. LAVP is specifically designed to overcome parking issues and provide intelligent transportation services in urban areas. LAVP responds to a parking call by providing the whole city's CP status to the vehicles.

4.1 *Big picture of LAVP*

LAVP provides the possibility of taking AVs from the drop-off spot to the selected CPs autonomously, as depicted in Fig. 3. Usually, drop-off spots are deployed near congested areas, like the city center, shopping mall, hospitals and stadiums, while CPs are deployed in less congested and remote areas, which are usually on the border line of the city. In LAVP, a driver may at any time request for a parking lot in remote CPs and drop off AVs at a selected drop-off spot. The Scheduling Center (SC) has an important role in scheduling parking operation and providing an optimized solution that reduces journey time, fuel consumption and parking fees. The process of LAVP starts with a parking request for an outbound trip, e.g., to the office or WP. It suggests an efficient selection of drop-off spots to the AV. It helps the driver in leaving the AV at the spot nearest to the WP.

4.2 *Communication signalings in LAVP*

In Fig. 4, the communication framework is shown for both inbound and outbound journeys, where vehicles and infrastructures are denoted by "X" in the V2X systems.

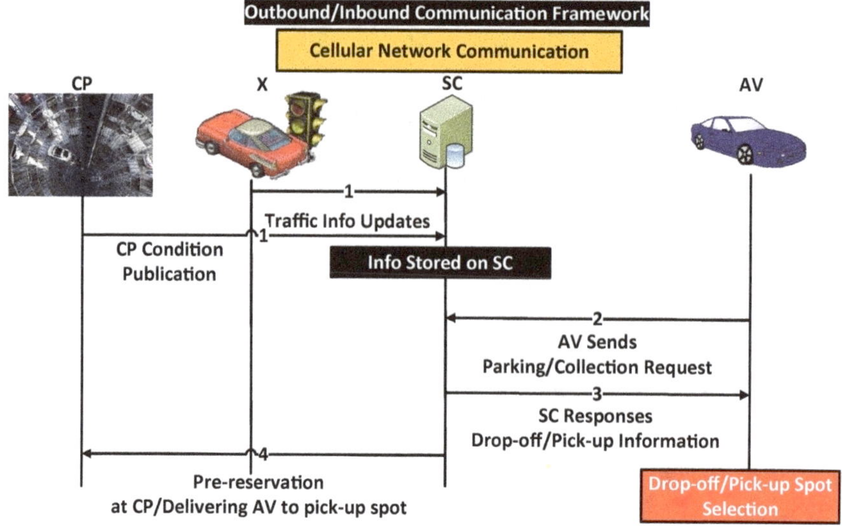

Fig. 4. The communication frameworks for outbound/inbound trips.

"Drop-off" spots for an inbound trip are regarded as "pick-up" spots. The "drop-offs/pick-ups" (D/P) and well-suited parking locations are scheduled by the SC keeping in view live traffic updates. SC considers the cost for the driver of the D/P and the cost incurred in delivering the AV to/from the CP.

The design presented in Fig. 4 is based on the cloud system, while mobile edge computing (MEC) [4] may be integrated in future. It will actually replace the operation of the SC. These edges collect traffic and vehicles' data and perform intelligent data calculations at their end. These results are then shared with other edges as well as the centralized cloud. An edge directly responds to AVs upon receiving a parking request as the local edge can be directly accessed by the AVs. These edges help in making decisions instantly and do not require handshakes/signaling.

4.3 *LAVP system cycle*

In Fig. 5, five stages of LAVP have been described.

Traveling Phase: In this phase, the AV is traveling in an urban city.

Drop-off for Outbound Trip: Moving toward the WP, when the AV is within a certain range of the WP, it sends a parking request to the SC.

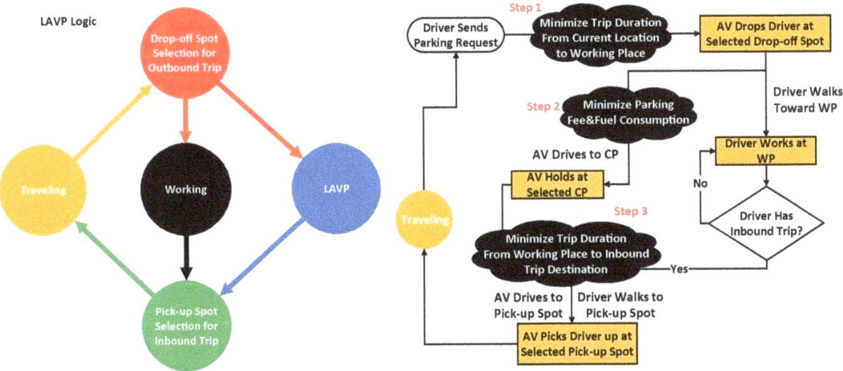

Fig. 5.   System cycle of LAVP procedure.

Working in Office and LAVP: As soon as the SC receives a parking request, it suggests a feasible drop-off spot to the driver. The driver then proceeds to the suggested drop-off spot, leaves the AV and starts walking toward the WP. In the meantime, the AV starts moving toward the CP and gets parked.

Pick-up for Inbound Trip: After working in the WP, the driver sends an inbound trip request to the SC. The SC schedules the AV delivery to the pick-up spot depending upon the time the driver will leave the WP and the time the driver will take from the WP to the pick-up spot. SC suggests the best available pick-up spot. The driver picks up the AV and then turns to the *driving phase* by driving toward the destination.

5   Scheduling Scheme of LAVP

5.1   *Problem definition*

Mathematically, transportation networks can be considered as multiple optimization problems. The network can be interpreted as follows: D_0 can be indicated as the start of a journey or some other point in the city, while D_T shows the target location or location where a person works, like the WP. In the transportation network, there are multiple factors that need consideration but, for instance, the goal is to minimize the the time spent from D_0 to D_T. Let us suppose x denotes the possible locations of the traffic network. Let $f(x)$ be the shortest path connecting D_0 and x and let $g(x)$ be the shortest path connecting x and D_T. Let the set of M drop-off points be

$$\{\mathrm{drop}_i\}_{i=1}^{M}. \tag{6.1}$$

To make it simpler, the speed of the AV, S_v, and that of the person walking, S_h, are kept constant. Hence, the first optimization problem is

$$x^* = \underset{x \in \{\text{drop}_i\}_{i=1}^M}{\arg\min} \left(\frac{f(x)}{S_v} + \frac{g(x)}{S_h} \right). \tag{6.2}$$

The main theme is to minimize the time spent on the way from D_0 to D_T when the AV is used. After dropping the person at x^*, the AV will autonomously move toward one of the predefined CPs in the remote area. The CPs' locations and vacant parking slots at time t as a set are given as follows:

$$\text{Cap}(t) := \{\text{park}_i\}_{i=1}^N. \tag{6.3}$$

The capacity of CPs should be updated regularly. Let the time when the vehicle drops the person at x^* be t_0 and let $l(t_0, x)$ denote the shortest distance between x^* and a CP $x \in \text{Cap}(t_0)$. The aim is to minimize the parking cost and expenses on the way to the CPs, namely,

$$\min_{x \in \text{Cap}(t_0)} \left(a \times \frac{l(t_0, x)}{s_v} + b(x) \times \omega \right), \tag{6.4}$$

where the cost of electricity or gas is denoted by a, while $b(x)$ denotes the parking cost for one hour in each CP. The total time of parking there is denoted by ω, which is assumed to be constant now. At time t, we consider

$$\min_{x \in \text{Cap}(t)} \left(a \times \frac{l(t, x)}{s_v} + b(x) \times \omega \right). \tag{6.5}$$

The $l(t, x)$ function connects the location of the vehicle at time t and the CP $x \in \text{Cap}(t)$ through the shortest distance. Although it is difficult to do it at time t, we may do it at a discrete time t and update the capacity information of CP, like once every 5 minutes.

Here, it is worth noting that if any reservation function is used for parking the AV, the optimization problem (6.4) may be considered to find out the best CP in $\text{Cap}(t_0)$ to confirm the reservation.

5.2 *Generic LAVP computation logic*

The computation logic in LAVP is developed to make an appropriate selection of the drop-off spot for an outbound trip and the pick-up spot selection for an inbound trip. It will minimize the trip duration with a trade-off between fuel consumption/parking cost. Here, $d/p \in \mathcal{D}$ is shown as a set of D/Ps, and $cp \in \mathcal{P}$ as the number of CPs in a network.

- **Step 1:** In the initial step, SC selects a drop-off spot, considering the minimization of traveling time. The selection of drop-off spot depends on the current location of the AV. It can be achieved by $\arg\min_{d \in \mathcal{D}}(\frac{D_{d,w}}{S_h} + \frac{D_{v,d}}{S_v})$, here $D_{d,w}$ is the distance[1] between a drop-off spot and the WP, while $D_{v,d}$ is considered as the distance between the drop-off spot and the current location of the AV as presented in Fig. 4. Besides, S_h (where h stands for human) and S_v are assumed as average walking and driving speeds, respectively.

- **Step 2:** It is the responsibility of the SC to determine a suitable CP for the AV left by the driver at a drop-off spot. Usually, CPs with vacant slots are considered for this process. In this step, fuel consumption for the return trip to the drop-off spot and the parking fee for which the AV will be parked are taken into consideration. In cases where the parking fees are the same, the CP with the shortest traveling distance will be selected. This can be achieved by calculating $\arg\min_{cp \in \mathcal{P}} D_{d,cp}$.

- **Step 3:** In step 3, the driver needs to collect AV from the pick-up spot. The SC selects a pick-up spot keeping in view the current location of the driver. A pick-up spot will be selected based on certain criteria like minimum travel time and traveling expense if the driver is using public transport toward the pick-up spot. It can be achieved by $\arg\min_{p \in \mathcal{D}}(\frac{D_{w,p}}{S_h} + \frac{D_{p,t}}{S_v})$, where the distance between the pick-up spot and the WP is presented by $D_{w,p}$, while $D_{p,t}$ is the distance between the pick-up spot and the inbound trip destination.

5.3 *Analysis*

5.3.1 *LAVP vs. benchmark*

An analysis has been provided to show the advantage of LAVP; here, the inbound and outbound trips are the same. We are taking only the inbound trip into account. There are no D/P spots involved in the benchmark. In fact, the driver needs to drive all the way to the CP, get their car parked and walk back toward the WP.

---

[1]Although we only illustrate the geometric linear distance for simplicity of presentation, in a case study, the actual path consisting of the segment coordinates of the road map is considered in calculating the distance.

In the case of LAVP, the outbound trip $T^{\text{out}}_{\text{lavp}}$ is given by

$$T^{\text{out}}_{\text{lavp}} = \frac{D_{v,d}}{S_v} + \frac{D_{d,w}}{S_h}, \tag{6.6}$$

and that for benchmark is given by

$$T^{\text{out}}_{bck} = \frac{D_{v,\text{cp}}}{S_v} + \frac{D_{\text{cp},w}}{S_h}. \tag{6.7}$$

As S_v is much larger when compared to S_h (e.g., 13.9 m/s vs. 1.5 m/s), mainly $D_{d,w}$ and $D_{\text{cp},w}$ dominate how advanced the LAVP is. In general, the deployment of D/P reflects the efficacy of LAVP, while large CPs are usually expected in the remote areas of each city, rather than near the city center as in the example given in Fig. 3. With a large number of drop-off spots, it is possible to find a $d \in \mathcal{D}$ to hold $D_{d,w} < D_{\text{cp},w}$.

5.3.2 *Convenience vs. fuel consumption*

Regarding the deployment of D/P, let us assume a simple case, with one CP, where $|\mathcal{P}| = 1$, and one WP. Particularly, $D_{d,w} = 0$ means the drop-off spot is co-located with WP. Here, a triangle is formed by $D_{v,d}$ (note that $D_{v,d} = D_{v,w}$, since $D_{d,w} = 0$), $D_{v,\text{cp}}$ and $D_{\text{cp},w}$. Obviously, we can obtain $D_{v,w} < (D_{v,\text{cp}} + D_{\text{cp},w})$ according to Euclidean geometry. Note that $D_{v,\text{cp}} + D_{\text{cp},w}$ is actually the traveling distance spent in the benchmark solution. This analysis provides an insight that the drop-off spot should normally be set close to the WP, which certainly follows the vision of the LAVP system to the benefit of users in terms of driving experience.

As the fuel consumption is proportional to the distance an AV travels, the distance $D_{v,d} + D_{d,\text{cp}}$ is traversed in case of LAVP. In the benchmark, $D_{v,\text{cp}}$ is traversed. Obviously, the LAVP will result in much fuel consumption as given in the above simple case, as there is only one CP. However, with more CPs built in the city, the fuel consumption of LAVP can be potentially reduced by diverting the AV toward a CP that is closer to the drop-off spot (as the CP selected in LAVP and the benchmark does not need to be the same).

6 Case Study

The case study is implemented under Opportunistic Network Environment (ONE) [13], a Java-based simulator originally used for DTN routing research. The default scenario with the $4500 \times 3400\,\text{m}^2$ area is shown as the

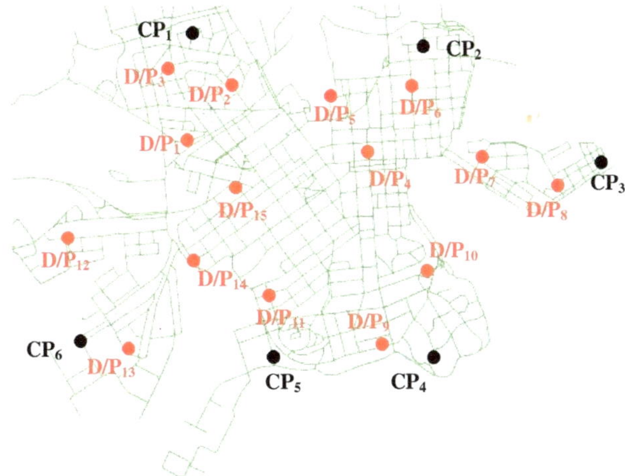

Fig. 6. The Helsinki city scenario for case study (6 CPs, 15 drop-off/pick-up spots).

down-town area of Helsinki City in Finland. A total of 300 AVs running at speeds in the range [30 $\sim$ 50] km/h are initialized in the network, as shown in Fig. 6. A total of six CPs in the remote area are deployed while 15 drop-off/pick-up spots (depicted as "D/P") in total are deployed in the main city center. By default, the time for drivers to start requesting for drop-off spots is 3,600 seconds while 7,200 seconds is set as the working period. The simulation runs for 12 hours. Here, the power demand (P) of an AV can be calculated in [24].

The duration experienced by an AV in outbound and inbound trips can be denoted by $\mathcal{H}$, while fuel consumption in (J) can be given by $\int_0^{\mathcal{H}} P$, when acceleration is enabled.

For the proposed LAVP, results are shown given the different deployment of D/P spots. In cases where there are 4 D/P spots, D/P$_4$, D/P$_{10}$, D/P$_{11}$ and D/P$_{15}$ are deployed. Ten D/P spots means D/Ps other than (D/P$_4$, D/P$_{10}$, D/P$_{11}$, D/P$_{14}$ and D/P$_{15}$) are considered. Fifteen D/P spots means all the D/Ps are deployed. For the purpose of a fair comparison, the parking fee of all CPs is set as the same, as the main interest is to compare LAVP with the benchmark on the basis of the following performance metrics:

- **Average Walking Duration (AWD):** The average period for drivers to move from drop-off spots (in LAVP)/CPs (in the benchmark) toward WPs for outbound trips, plus that for inbound trips.

- **Average Trip Duration (ATD):** The average time that drivers experience for their trips. This is measured from the time they request drop-off until the time they reach the WPs, for outbound trips. For inbound trips, this includes the time taken for drivers to reach the pick-up spots from WPs up to the time when they reach the inbound trip destinations. The result accumulates the ATD of these two periods.
- **Total Fuel Consumption (TFC):** This refers to all the AVs, including both outbound and inbound trips.

The results are presented with a normalized value. In Fig. 7, the outcome of deploying a large number of D/P spots can be observed, which helps the driver in reducing the AWT.

Besides, the ATP is reduced, as certain D/P spots can be selected close to the drivers' WPs. In particular, in case of 4 D/P spots, both LAVP and the benchmark achieve a close AWD and ATD. Here, although the ATD is

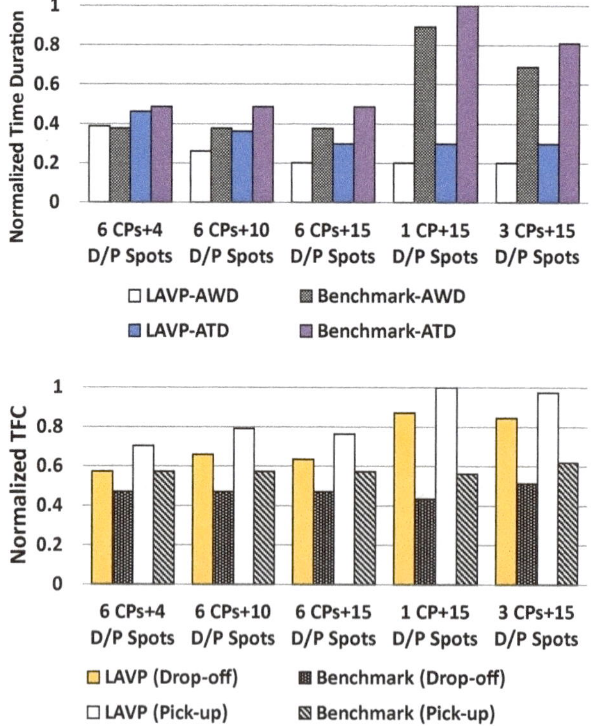

Fig. 7. Evaluation results.

reduced compared to the benchmark, the AWT cannot get reduced due to the limited number of deployed D/P spots (so it does not benefit the drivers to walk to their WPs). The results demonstrate that the proposed LAVP has the capability of improving the user QoE (in terms of shorter journey time), and the more the deployment of D/P spots, the more benefits it achieves. If we are only keeping CP_1 in the network, the benchmark suffers from the highest AWT and ATD, as shown in the fundamental analysis in Section 5.3, while, with three CPs (CP_1, CP_3 and CP_6 included), the benchmark obtains lower AWT and ATD.

In case of LAVP, the TFC is increased when more D/P spots are deployed. This is because AVs would first drive toward the D/P spots (which primarily benefit the drivers) and later head to the CPs/inbound trip destinations. In particular, deploying four D/P spots enables one to uniformly cover the needs around the central city, compared to the 10 D/P spots case. Therefore, the latter case achieves a higher TFC. Further deploying D/P spots from 10 to 15, TFC is reduced as AVs can find convenient D/P spots for drivers working around the central city. If we increase one CP to three CPs, LAVP obtains reduced TFC compared to the benchmark.

7 Conclusion

This chapter discussed the LAVP system, which relies on support from both ICT and autonomous driving. With a number of deployed D/P spots and pretrained track routes to/from remote CPs, LAVP reduces the walking time taken by the drivers to walk between the WPs and the total out/inbound trip duration. With a trade-off at fuel consumption, the highly improved QoE makes the LAVP promising in the future ITSs. Our ongoing work will focus on reasonable pricing and reservation system.

References

[1] Ali, Q. E., Ahmad, N., Malik, A. H., Ali, G., Asif, M., Khalid, M., & Cao, Y. (2018). SPATA: Strong pseudonym-based authentication in intelligent transport system. *IEEE Access*, 6, 79114–79128.

[2] An, K., Choi, J., & Kwak, D. (2011). Automatic valet parking system incorporating a nomadic device and parking servers. In *IEEE International Conference on Consumer Electronics (ICCE)*, pp. 111–112, doi:10.1109/ICCE.2011.5722489.

[3] Arshad, M., Ullah, Z., Khalid, M., Ahmad, N., Khalid, W., Shahwar, D., & Cao, Y. (2018). Beacon trust management system and fake data detection

in vehicular ad-hoc networks. *IET Intelligent Transport Systems*, 13(5), 780–788.

[4] Beck, M. T., Werner, M., Feld, S., & Schimper, S. (2014). Mobile edge computing: A taxonomy. In *Proceedings of the Sixth International Conference on Advances in Future Internet*, Citeseer.

[5] BNTS (2016). British national travel survey. https://www.gov.uk/government/statistics/national-travel-survey-2016 [Accessed November, 2018].

[6] Broggi, A., Cardarelli, E., Cattani, S., Medici, P., & Sabbatelli, M. (2014). Vehicle detection for autonomous parking using a soft-cascade adaboost classifier. In *IEEE Intelligent Vehicles Symposium Proceedings*, pp. 912–917, doi:10.1109/IVS.2014.6856490.

[7] Cao, Y., Ahmad, N., Kaiwartya, O., Puturs, G., & Khalid, M. (2018). Intelligent transportation systems enabled ICT framework for electric vehicle charging in smart city. In *Handbook of Smart Cities*, Springer, pp. 311–330.

[8] Chirca, M., Chapuis, R., & Lenain, R. (2015). Autonomous valet parking system architecture. In *IEEE 18th International Conference on Intelligent Transportation Systems*, pp. 2619–2624, doi:10.1109/ITSC.2015.421.

[9] Department for Transport (2016). Transport statistics Great Britain, 2016. Technical Report Accessed [24th April 2018].

[10] Gasser, T. M. *et al.* (2012). Legal consequences of an increase of vehicle automation. Technical Report, The Federal Highway Research Institute, Magazine F 83.

[11] Geng, Y., & Cassandras, C. G. (2013). New smart parking system based on resource allocation and reservations. *IEEE Transactions on Intelligent Transportation Systems*, 14(3), 1129–1139.

[12] JustPark (2016). Justpark survey report. https://www.justpark.com [Accessed 12th November 2018].

[13] Keränen, A., Ott, J., & Kärkkäinen, T. (2009). The ONE simulator for DTN protocol evaluation. In *ICST SIMUTools '09*, Rome, Italy.

[14] Khalid, M., Cao, Y., Arshad, M., Khalid, W., & Ahmad, N. (2017). Routing challenges and associated protocols in acoustic communication. In *Magnetic Communications: From Theory to Practice*, CRC Press, pp. 109–126.

[15] Khalid, M., Cao, Y., Aslam, N., Suthaputchakun, C., Arshad, M., & Khalid, W. (2018). Optimized pricing & scheduling model for long range autonomous valet parking. In *2018 International Conference on Frontiers of Information Technology (FIT)*, IEEE, pp. 65–70.

[16] Khalid, M., Cao, Y., Zhang, X., Han, C., Peng, L., Aslam, N., & Ahmad, N. (2018). Towards autonomy: Cost-effective scheduling for long-range autonomous valet parking (LAVP). In *IEEE Wireless Communication & Networking Conference*, IEEE Barcelona, Spain.

[17] Khalid, M., Ullah, Z., Ahmad, N., Arshad, M., Jan, B., Cao, Y., & Adnan, A. (2017). A survey of routing issues and associated protocols in underwater wireless sensor networks. *Journal of Sensors*, 2017.

[18] Khalid, M., Ullah, Z., Ahmad, N., Khan, H., Cruickshank, H. S., & Khan, O. U. (2017). A comparative simulation based analysis of location based

routing protocols in underwater wireless sensor networks. In *IEEE Workshop on Recent Trends in Telecommunications Research (RTTR)* Palmerston North, New Zealand, pp. 1–5.

[19] Khalid, W., Ullah, Z., Ahmed, N., Cao, Y., Khalid, M., Arshad, M., Ahmad, F., & Cruickshank, H. (2018). A taxonomy on misbehaving nodes in delay tolerant networks. *Computers & Security*, 77, 442–471.

[20] Kotb, A. O., Shen, Y.-C., Zhu, X., & Huang, Y. (2016). Iparkera new smart car-parking system based on dynamic resource allocation and pricing. *IEEE Transactions on Intelligent Transportation Systems*, 17(9), 2637–2647.

[21] Lam, A. Y., Leung, Y.-W., & Chu, X. (2016). Autonomous-vehicle public transportation system: scheduling and admission control. *IEEE Transactions on Intelligent Transportation Systems*, 17(5), 1210–1226.

[22] Min, K., Choi, J., Kim, H., & Myung, H. (2012). Design and implementation of path generation algorithm for controlling autonomous driving and parking. In *12th International Conference on Control, Automation and Systems (ICCAS)*, Je Ju Island, South Korea, pp. 956–959.

[23] Min, K.-W., & Choi, J.-D. (2013). Design and implementation of autonomous vehicle valet parking system. In *16th International IEEE Conference on Intelligent Transportation Systems-(ITSC)*, The Hague, Netherlands, pp. 2082–2087.

[24] Nam, E. K., & Giannelli, R. (2013). *Fuel Consumption Modeling of Conventional and Advanced Technology Vehicles in the Physical Emission Rate Estimator (PERE)* (BiblioGov).

[25] Ni, J., Zhang, K., Yu, Y., Lin, X., & Shen, X. S. (2018). Privacy-preserving smart parking navigation supporting efficient driving guidance retrieval. *IEEE Transactions on Vehicular Technology*, IEEE, pp. 6504–6517, doi:10.1109/TVT.2018.2805759.

Chapter 7

Mobile Data Offloading for Smart Mobility in a Heterogeneous Network

Tong Wang[*,‡], Xibo Wang[*], Guang Xin Yang[*] and Yue Cao[†,§]
*College of Information and Communication Engineering,
Harbin Engineering University, Harbin 150001, P. R. China
†School of Computing and Communication,
Lancaster University, InfoLab21, South Dr, Lancaster University,
Bailrigg, Lancaster LA1 4WA, UK
‡wangtong@hrbeu.edu.cn
§yue.cao@lancaster.ac.uk

1 Introduction

Over the last couple of years, an increasing number of smart mobile devices have become available in the market, such as smartphones, tablets and laptops. The interest in using intelligent terminals for entertainment and resource-consuming services is one of the primary factors contributing to the global mobile traffic growth. According to the Cisco Visual Network Index, global mobile data traffic grew to 63% in 2016, reaching 7.2 exabytes per month at the end of 2016, up from 4.4 exabytes per month at the end of 2015 (1 exabyte is equivalent to 1 billion gigabytes, and 1,000 petabytes). Additionally, Cisco also forecasts that mobile data traffic will grow at a compound annual growth rate (CAGR) of 47% from 2016 to 2021 and reach 49.0 exabytes per month by 2021 [21].

Up to now, there already exist several major methods to alleviate the congestion and make the network more efficient. Apparently, the most straightforward one is to install more base stations per area, so that the capacity of the network is increased. As it is known to all, upgrading the cellular network to the next generation such as 5G [7] is another promising way. For example, in a novel 5G scenario, the vehicle nodes would download interesting media quickly [11, 29]. Generally, the network conditions can be improved by building extra infrastructures. However, on the one hand, it is costly and it will inevitably take a long period to upgrade the hardware equipment. On the other hand, it is uncertain whether it can keep pace with the explosively increasing mobile data [34]. Additionally, cellular operators have attempted to formulate rational pricing strategies to limit customers' demand for data usage. For example, Ha *et al.* divided a day into peak hours and off-peak hours [27]. Although such strategies successfully transfer a large quantity of mobile data from peak hours to off-peak hours, they actually degrade the user experience and affect user satisfaction.

Given these negative factors, we turn our attention to one solution, mobile data offloading, which recently has been attracting increasing interest from the research community. Although the overall radio spectrum is sufficient, only a small portion can be employed by the operators. Fortunately, offloading plays a significant part in saving cellular infrastructure bandwidth and makes the spectrum more efficient [15]. Moreover, the increasing maturity of communication interface technology, such as multiple interfaces and high speed, makes offloading feasible [3, 46]. Meanwhile, many mobile data offloading strategies have been proposed, including Wi-Fi access point (AP) [2], femtocell [16, 20, 62] and opportunistic communication [43].

The focus of this chapter is to provide a summary of existing mobile data offloading strategies in detail, by giving a comprehensive explanation for beginners. The major contributions are three-fold as given below:

- To categorize the existing mobile data offloading strategies into infrastructure-based and infrastructure-less branches.
- To provide an up-to-date review on previously published works with a discussion on the advantages and disadvantages.
- To discuss the performance metrics and challenges associated with offloading.

The remainder of this chapter is organized as follows. In Section 2, in terms of the role that infrastructure plays, we present a classification of the existing offloading schemes: infrastructure-based and infrastructureless. In Section 3, we propose an overview with a detailed analysis for each category. A variety of performance metrics proposed in the previous literature are introduced in Section 4. Finally, we discuss the challenges that may occur in the near future and conclude this chapter in Sections 5 and 6, respectively.

2 Review Taxonomies

Up to now, a great variety of offloading trials have been implemented in the academic and industrial community. Some major worldwide operators such as AT&T, T-Mobile and Verizon have installed more wireless APs in their networks to promote mobile data offloading. A few works such as [25, 51] have summarized and categorized the existing mobile data offloading techniques. Based on the type of the network frequency resources, Elhami *et al.* divided existing mobile data offloading strategies into three following categories:

- Using unlicensed bands as network resources.
- Using licensed bands as network resources.
- Using a combination of unlicensed and licensed bands [25].

They categorize the offloading methods as femtocell technology in licensed bands, Wi-Fi APs in the unlicensed bands, and device-to-device and opportunistic communications in either licensed or unlicensed bands.

In other papers [51], depending on the time sensitivity of mobile data, Aijaz *et al.* [2] divided the existing schemes into two categories: non-delayed offloading and delayed offloading as follows:

- In the case of non-delayed offloading, no extra latency is allowed in the whole offloading process. For example, it is essentially meaningless for a live video or audio file to be streamed once it is attached to an additional latency. Nobody will be interested in watching a basketball game after the result of the game is revealed.
- In contrast to non-delay offloading, delayed offloading is not associated with strict delivery delay constraints in which users can patiently wait for the data content until the deadline is reached.

The delay-tolerance feature ensures the content is delivered through multiple methods. In general, most of the users can readily accept this condition, e.g., that their acquired music or movies are received with a tolerable delay. If they fail to download the content while they are in a congested cellular network, users may prefer to obtain the content through a Wi-Fi AP in proximity. In addition, it is also a terrific option when receiving cached content shared by other peers.

In this section, we primarily present a comprehensive classification of the existing offloading strategies and further provide our understanding from another perspective. Throughout the study of the existing literature, we found that some offloading methods require the assistance of infrastructure, while others do not. Hence, we briefly classify the currently available offloading schemes into infrastructure-based offloading and infrastructure-less offloading. In Section 2.1, we will introduce more specific details of the offloading strategies from comprehensive perspectives.

2.1 *Infrastructure-based offloading*

In infrastructure-based offloading solutions, some infrastructures with communication, computing and storage capabilities are deployed to assist in data transmission in which Wi-Fi APs and femtocell (as two kinds of efficient complementary access networks) play important roles in guaranteeing the mobile data being transferred from the cellular network. These two main offloading technologies are discussed as follows.

Wi-Fi stands for wireless fidelity, which is a wireless connectivity solution based on IEEE 802.11 standards [2]. It was designed for getting internet access in an indoor environment with a data rate of 1 Mbps. In 2003, IEEE developed a new version, 802.11g, raising the data rate to 54Mbps. IEEE 802.11n, presented in 2007, efficiently combines MIMO and OFDM technologies, thereby improving the wireless communication quality; thus, the data rate was up to 600 Mbps [47]. What attracts cellular operators most is that Wi-Fi operates on the free unlicensed frequency bands (2.4 GHz UHF and 5 GHz SHF) without causing interference to the cellular network, which is an economical solution to expand the network capacity. The architecture of the Wi-Fi network is shown in Fig. 1; the data are transparently transferred to the Wi-Fi network when they are in Wi-Fi coverage, completely bypassing the cellular core network for data services.

Wi-Fi offloading possesses a high data transmission rate and has low requirements of devices [32]. Therefore, it is regarded as one of most promising techniques to process data increasing explosively in the cellular networks. As a natural solution to mobile data offloading, Wi-Fi has limited coverage due to its stationary position and finite transmit power [63]. Although most Wi-Fi works in indoor environments (e.g., office, airport and train station), operators are continuously deploying outdoor Wi-Fi APs in urban areas to meet the demands of ubiquitous access.

Another infrastructure-based solution for offloading is femtocell which is also known as home base station. Unlike Wi-Fi networks that operate in

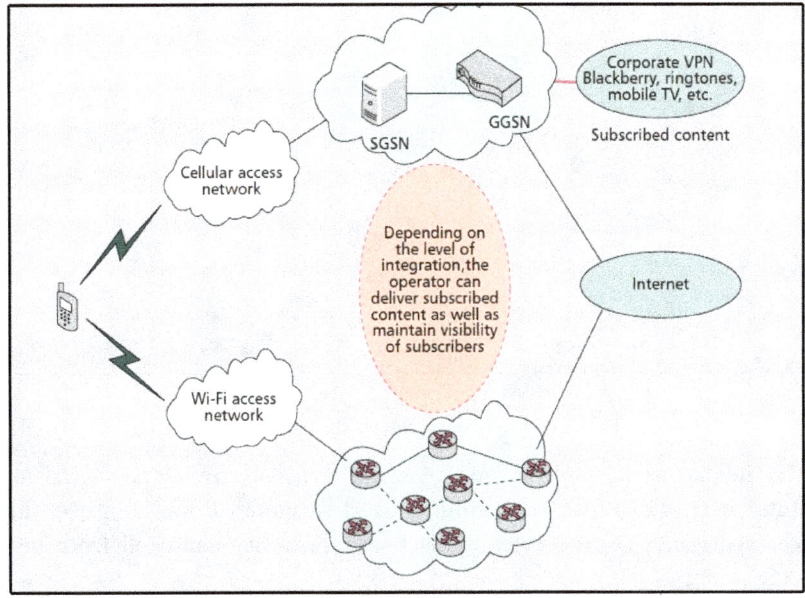

Fig. 1. Mobile data offloading via Wi-Fi network.

an unlicensed spectrum in which radio interference is not actively managed, femtocell networks will operate in the licensed spectrum. In a wider context, it is considered as a base station with limited coverage, and low power is employed by indoor users to get a better service. Similar to Wi-Fi AP, its transmitting power is about 10–100 mW with the capacity of 4–6 users. As shown in Fig. 2, the user-installed device communicates with the cellular network over a broadband connection such as a digital subscriber line (DSL), a cable modem or a separate radio frequency (RF) backhaul channel [14]. Previous studies have shown that more than 50% of all voice calls and more than 70% of data usage occurs in an indoor environment such as a home or an office. Therefore, as a potential method of offloading, femtocell can shift a large amount of traffic originating in the macrocellular network.

There are several benefits in using femtocell to alleviate data congestion in cellular networks for both operators and users. First, when compared with installing a macrocell tower, it is more convenient and less expensive for operators to deploy this home base station. Moreover, users can enjoy better services in terms of seamless experience and higher bit rate. Last but not least, due to not managing another interface and consuming energy to scan the available APs, femtocell prolongs the battery life of devices [64].

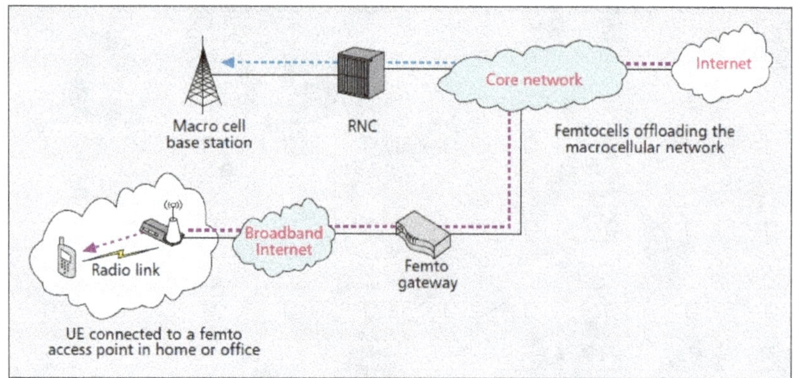

Fig. 2. Mobile data offloading via femtocell.

Nevertheless, as a result of operating on the same frequency band as a cellular network, interference from a nearby macrocell and femtocell may affect the quantity of service. This factor prevents femtocell from being widely adopted.

2.2 *Infrastructureless offloading*

The positive impact of the increasing popularity of smart mobile devices is two-fold. On the one hand, there is a corresponding increase in the amount of infrastructures deployed in infrastructure-based solutions. On the other hand, smart mobile devices equipped with several alternative communication options make it possible to transmit data directly between mobile users, without any need for an infrastructure. Thus, some operators propose infrastructureless solutions.

Here, we mainly discuss offloading mobile data through opportunistic networks or delay tolerance networks (DTNs). Both the terms, opportunistic networks and DTNs, refer to the use of opportunistic contacts of mobile terminals to finalize communication, as shown in Fig. 3. Opportunistic communication technology makes it possible for mobile terminals to communicate with other mobile terminals directly without the involvement of APs or base stations [36]. Recently, the advanced interface technologies of intelligent terminals such as cellular, bluetooth and Wi-Fi direct interface have boosted the development of opportunistic communication. Such a technique ensures that users access both the cellular network and the local network at the same time.

Because most of the bluetooth and Wi-Fi direct technologies are operated on the 2.4 G unlicensed frequency band without additional equipment,

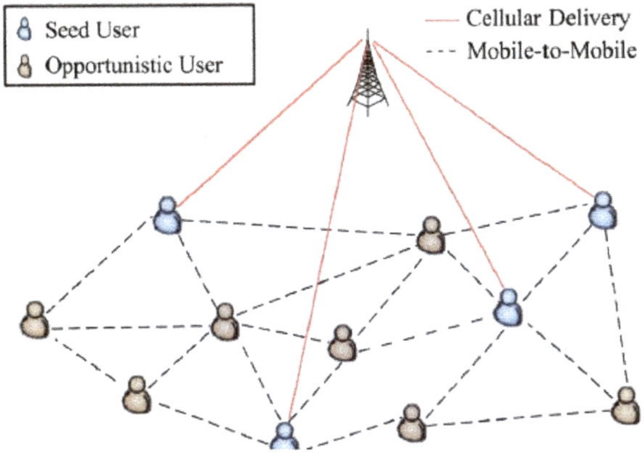

Fig. 3.   Mobile data offloading via opportunistic communication.

using opportunistic communication as an offloading method means an extremely low monetary cost for operators. Additionally, intelligent terminals with a long battery life, accurate GPS and large storage make offloading through opportunistic communication more competent. Nowadays, a large amount of mobile data such as multiple-media newspapers, popular videos and music are of common interest among a set of users, and most of these contents are not associated with strict real-time constraints. Benefiting from this delay-tolerant characteristic, operators can only transfer the content to a small number of users. The content is going to be further delivered to all subscribers by those selected users through opportunistic communication.

It is essential for any offloading scheme to select mobile users with high potential to further spread the content. The existing offloading strategies can be briefly classified into two main types: topology awareness and social status awareness. The former makes offloading decisions based on real-time topology information, while the latter puts more focus on the social status of mobile users. It is obvious that the connectivity is immediately established once pairwise users come into the communication range of each other [56].

The topology information here exactly refers to the connectivity between numerous users. In general, users can send contents to their current neighbors who have good qualifications for dissemination. For instance, only a number of neighbors beyond a threshold can trigger the dissemination process. More topology-awareness algorithms such as epidemic, direct delivery and spray and wait (SnW), which are equally efficient in offloading, are described in [54, 65].

The social status-awareness offloading strategies exploit influence on social interaction among numerous mobile users to identify the potential helpers. The social status may include the user's mobility pattern, the interacting time, the real-time location and the activity [6]. Universally, a user who has more active mobility and encounters a large number of peers will be regarded as a potential helper. This indicates that helpers, behaving as a source or a relay, are more likely to successfully finalize the delivery. Similarly, a long interacting time means a stable connectivity between users which can achieve a large size file transmission.

Moreover, the encounter times between two users are also used to identify the degree of intimacy between the two users. Imagine that Lily has cached the content required by Bob, but Lily and Bob are far apart. Fortunately, Lily encounters Bob's intimate fellow Smith. So, it is very likely that Smith will send the content to Bob eventually after he receives it from Lily. More social status-awareness routing algorithms such as PRoPHET, MaxProp and BUBBLE Rap are described in [8, 30, 39]. Although having the lowest cost, this offloading approach also has some limitations. The classification of existing offloading strategies as well as their benefits and shortcomings are listed in Table 1.

Table 1. Classification of offloading strategies as well as their benefits and strategies.

Classification		References	Benefits	Shortcomings
Infrastructure-based offloading	Wi-Fi	[4, 17–20, 23, 24, 35, 45, 52]	1. Less expensive 2. No interference 3. High data rate and reliability	Limited coverage and mobility
	Femtocell	[1, 9, 12, 48, 66]	1. Easy deployment 2. Prolonged device battery life 3. Capacity gain from higher SINR	Interference from nearby macrocells and femtocells
Infrastructureless offloading	Topology	[13, 37, 42, 49, 50, 57, 59]	1. No monetary cost 2. No extra equipment 3. High efficiency dynamic environment	1. Consumption of battery and storage 2. Privacy and security problem
	Social status	[5, 22, 26, 28, 30, 31, 33, 38, 40, 41, 58, 61]		

3 Review on Existing Offloading Solution

Following the previous classification, a review on previous works in each category is discussed in this section.

3.1 *Infrastructure-based offloading solutions*

3.1.1 *Mobile data offloading via Wi-Fi*

As far as mobile data offloading is concerned, utilizing Wi-Fi networks can be the most promising solution as it can offer an immediate remedy to this problem. There are many Wi-Fi APs available at many user locations such as homes, shops and universities. Due to its various advantages and its promising future, Wi-Fi offloading has become a hot research topic and has attracted the attention of many researchers.

Ristanovic *et al.* came up with the two algorithms — MixZones and HotZones — for offloading mobile data from 3G networks [52]. Hot Zone is a Wi-Fi-based offloading scheme and refers to a cell covered by Wi-Fi APs. With the help of user mobility profiles (UMPs), the operator ranks the cells for users according to their average visiting number in a day. If a cell has a large visiting number, that implies that the user has obtained permission to access the Wi-Fi (considering many Wi-Fis are not provided for free). Balasubramanian *et al.* proposed Wiffler, a system to improve cellular network capacity through a Wi-Fi AP deployed in an urban environment [4]. An easy approach is to transfer data with a certain delay via the Wi-Fi interface, until a Wi-Fi AP is available. As a result, the proposed Wiffler has the ability to offload almost 50% of the total 3G traffic with an average delay of 60 seconds.

With the development of intelligent traffic, Wi-Fi offloading in vehicular environments has attracted attention [17–19]. Generally, vehicles signal to nearby WiFi APs when traveling along a road, so that the cellular traffic can be delivered to vehicles. In [18], the authors have provided an overview of the mobile data WiFi offloading in vehicular communication environments and discussed its unique characterizations, effectiveness, technical challenges, existing solutions and future research directions. In [19], the authors have proposed an approach to predict Wi-Fi offloading potential and access cost in vehicular environments and introduced auction game-based and congestion game-based offloading mechanisms for vehicular users to effectively offload the cellular traffic through the carrier Wi-Fi network. In [17], the authors have theoretically investigated the performance of vehicular Wi-Fi offloading, modeled the arrivals and fulfillments of the data services of a vehicular user as an M/tt/1/K queue, derived the probability distribution of the effective service time and analyzed the average service delay and offloading effectiveness by using the queueing analysis.

On the contrary, Cheung and Huang point out that most of the previous works mainly apply as little cellular data as possible, ignoring the quality of service (QoS) of the applications [20]. Therefore, they pay more attention to the usage-based pricing, aiming to trade-off the user's cost and the QoS associated with the transmission deadline. They first analyze a delay-aware Wi-Fi offloading problem as a finite-horizon sequential decision problem. The delay-aware Wi-Fi offloading and network selection (DAWN) is presented as a solution to obtain an optimal QoS under the user's cost constraint. Generally, in the absence of a threshold structure, a sequential decision problem is hard to cope with [44]. Thus, considering the optimal policy has a threshold structure in deadline and file size, they further present a monotone approximation DAWN algorithm to solve more general offloading issues. The insight shows that it is not an optimal choice to offload via Wi-Fi when the deadline is tight and the Wi-Fi network is heavily loaded. On the contrary, users can benefit a lot from delay-aware off-loading when there is a long period before the deadline. It has been discovered that persistent effort should be shifted from static offloading schemes to dynamic offloading in order to adjust to a dynamic network condition.

Energy consumption is another key constraint that affects the performance of the offloading scheme. Yoshihisa *et al.* introduced a predict-based method maximizing the offloading data fraction with the constraint of energy consumption taken into account [45]. Here, by tracking the user's browsing print, the method takes sometime to develop an in-depth knowledge of the user's browsing habits and their frequently accessed Web pages. About 11% of data can be offloaded with a lower energy consumption than that required in the cellular network. If the predicted accuracy cannot be strictly guaranteed, a considerable amount of storage and energy is going to be wasted to cache the content which the user will hardly request later. Besides, another energy-aware offloading algorithm is proposed in [24] in order to improve the energy efficiency. Throughout the prediction, the traffic is offloaded to a Wi-Fi AP if the energy consumed using the Wi-Fi transmission minus the energy consumed in transmitting the same data volume using the cellular network is larger than a predefined threshold.

The above works were mostly tested in the virtual simulation platform, whereas some researchers tested Wi-Fi offloading schemes in a realistic scenario. Lee *et al.* presented one of the first quantitative studies on the performance of 3G cellular data offloading via Wi-Fi APs in South Korea [35]. About 100 volunteers implemented an iPhone application to track their Wi-Fi connectivity for over 2 weeks. From the data collected by the application, the authors found that 70% of their time is spent in a Wi-Fi environment and these users will come back within 40 minutes every time they leave the Wi-Fi coverage. Supported by a real dataset recording the mobility of 536 taxis in San Francisco for 30 days, Dimatteo *et al.* proposed an offloading architecture named the MADNet architecture [23]. The user's data requests have been classified into download requests (DR) and upload requests (UR). For DR, each time a data item is required, the user states a "final location" and the time when they will reach there. When the time expires, the item will not attract the user's interest anymore. For UR, a user records and uploads content at the time he enters the vicinity of a point of interest. The trace-driven experiment shows 50% of mobile data for both download and upload can be offloaded from a 3G network with hundreds of Wi-Fi APs.

3.1.2 *Mobile data offloading via femtocell*

Compared with Wi-Fi, femtocells are attractive to operators as they provide improvement in both coverage and capacity, especially for indoors. However, because femtocells operate on the same frequency band as a cellular network, interference from a nearby macrocell and femtocell may affect the quality of service. Thus, Zhu *et al.* proposed an approach to allocate appropriate radio resources for non-CSG UE data packet transmission in order to mitigate the effect of macro-femto interference without the need of a true handover [66]. The key benefit of this proposed mechanism is to achieve highly efficient mobility by enabling a CSG femtocell base station to relay data without handover and access to the CSG femtocell. This avoids potential interference and saves radio resource and signaling load on the network.

In addition, dense deployment of femtocells may cause interference between the two nearby femtocells. With the consideration of random and uncontrolled interference, Chandar *et al.* proposed an analytical model to estimate the capacity enhancement of the OFDMA-based co-channel macrocell–femtocell networks [12]. Based on a mathematical framework structured by the expected distance of the interfering femtocell, system load and throughput of both macrocell–femtocell users, the authors find that there is a positive correlation between the maximum user number accommodated by femtocells and the femtocell density. Additionally, transmitter power also plays a significant role in the average throughput, which is determined by the macrocell load, femtocell amount and traffic requirement determined. High transmitter power can also lead to a maximum offloading gain.

Considering the interference problem, some researchers explored the combination of femtocell and Wi-Fi networks. Ahn *et al.* proposed a scheme for traffic offloading between femtocell and Wi-Fi networks utilizing software-defined networking (SDN) technology [1]. The proposed scheme utilizes the property of the femtocell, wherein data traffic to the operator's core network is transmitted via the user's broadband network. SDN technology allows a terminal to maintain existing sessions after offloading through a centralized control of the SDN-based equipment. Offloading target selection schemes based on available bandwidth estimation and an association control mechanism can reduce the femtocell load while ensuring quality of service (QoS) in terms of throughput: it rarely causes performance degradation and connection losses. Experimental results on an actual test-bed showed that the proposed offloading scheme provides seamless connectivity and reduces the femtocell load by up to 46% with the aid of the proposed target selection scheme, while ensuring QoS after offloading. Mahmoud *et al.* presented a generic framework combining mobile femtocells with Wi-Fi networks (MFW) [48]. In contrast to traditional femtocells installed in an indoor environment, mobile femtocells are installed in public vehicles such as city buses and streetcars to offer internet access to mobile terminals. This kind of mobile femtocell is connected to a Wi-Fi transmitter on the vehicle roof, and the Wi-Fi transmitter connected to urban Wi-Fi APs offers a backhaul for the mobile femtocell. The result reveals that, with a maximum capacity of 40 users, MFW can offload up to 50% of data traffic originating in a cellular network.

Furthermore, Calin *et al.* provided an analysis on how to deploy femtocells to maximize the offloading gains in macrocell networks [9]. Because of propagation losses, indoor users may suffer from a severe degradation in signal-to-noise ratio (SNR). However, supposing indoor users are served by a femtocell, both the QoS of the user and the capacity of the macrocell are improved simultaneously. Throughout the analysis of different SNR values, they find that the offloading gain varies with macrocell transmit power in a negative correlation. A higher transmit power results in a higher SNR, thus more users can be provided service via the macrocell. Meanwhile, the mutual interference between a macrocell and a femtocell becomes more severe, when a high transmit power is configured. Concerning femtocell offloading, how to appropriately deploy femtocells to limit interference is still worth investigating.

3.2 *Infrastructureless offloading solutions*

Offloading traffic through opportunistic communications has been recently proposed as a way to relieve the current overload of cellular networks. Therefore, an increasing body of work is investigating the use of infrastructure-free opportunistic networking to complement existing cellular infrastructure. A number of strategies can be implemented to disseminate content among mobile nodes.

3.2.1 *Topology awareness*

Chandrasekaran *et al.* proposed an information-centric network (ICN)-based offloading architecture that propagates the content already being cached by helpers to other subscribers in proximity through D2D communication [13]. Such single-hop schemes degrade the network overhead. But in a sparse environment, this scheme is going to be less efficient. Additionally, when several helpers can serve one requester at the same time, avoiding transmission conflict is not discussed. The work in [59] presents a subscribe-and-send (SaS) architecture and an opportunistic forwarding protocol called HPRO to determine message routing. Different from most offloading solutions, it is the mobile user, instead of the service provider, who decides whether to deliver the content requested by others. Content with common interest is transferred to a denser region, so that many subscribers can profit. Nevertheless, users in a sparse area located in the opposite direction of the data stream may fail to obtain content through this offloading scheme.

In fact, identifying target users is an NP-hard problem. Li *et al.* [37] established a complicated mathematical framework to tackle the offloading problem. However, unlike previous works, they considered that (1) users have diverse subscribing preferences, (2) the content has various delay sensitizers and size, and (3) the storage capacity of the terminal is limited. Through a strict mathematical derivation, they translated this offloading problem into a submodular maximization under several linear constraints (including storage capacity, users' preference and heterogeneous content). Three suboptimal algorithms called greedy algorithm (GA), approximation algorithm (AA) and homogeneous algorithm (HA) are proposed to efficiently respond to different network conditions. GA is devised to offload mobile traffic in a generic scenario, while AA performs better than GA and HA in the situation that content in a DTN-based network is configured with a short lifetime. Considering some works do not capture the need to reward seed users, Filippo *et al.* included a rewarding cost in the design of the opportunistic offloading strategy [50]. They designed a global strategy to select which nodes act as seeders and which ones as leechers in order to reduce the total dissemination cost. In the above works, mobile users are assumed to be volunteers, who selflessly deliver content to every other user in their proximity. However, practical users are selfish in the delay-tolerant network formed by device-to-device (D2D) communications. Wang *et al.* took user selfishness into consideration and proposed a network formation game to capture the dynamic characteristics of selfish behaviors [57]. Simulation results show that user selfishness highly degrades data offloading efficiency compared with ideal volunteer users.

Due to the delay-tolerant nature of some multimedia content, mobile operators can choose a certain number of users as target users to receive the initial content. These selected users will further disseminate the content to all subscribers in the network through opportunistic contacts. Finally, to ensure 100% delivery, the operator will directly send the content to the users who are unlikely to fetch it (within the utmost delay they can tolerate). In [42], Mayer *et al.* proposed a routing solution for the offloading of message exchange between end users where the protocol initially attempts to transmit messages through opportunistic communications and switches to the infrastructure-based network only when the probability of delivering the message within the deadline decreases.

Although opportunistic communication between mobile users is a promising alternative in mobile data offloading, designing an efficient solution is not an easy job. Rebecchi *et al.* proposed and designed a framework called Derivative Re-injection to Offload Data (DROiD) in which the core mechanism is to use the infrastructure as little as possible [49].

There is no doubt that this framework will be very efficient in a highly dynamic environment.

3.2.2 *Social status-awareness*

Obviously, how to wisely select the initial users who can maximize offloading gains is a key issue. With the social status of mobile users in mind, Han *et al.* presented three algorithms — random, greedy and heuristic — to investigate the optimal target-user selection [28]. Previous works have shown that there is a regularity in human mobility [26]. With this in mind, the heuristic algorithm selects target users by considering historical mobility. The heuristic outperforms the other two algorithms with over 73% of the total traffic offloaded.

However, tracking and recording the mobility of numerous mobile users consumes a lot of resources. Furthermore, any dramatically changed mobility pattern inevitably leads to performance degradation. Wang *et al.* proposed the TOSS framework to offload mobile cellular traffic for social network services via opportunistic sharing [58]. They discuss methods to choose the appropriate initial seeds, depending on their spreading impact in the SNS and their mobility impact in the offline mobile social networks (MSN). Trace-driven evaluation reveals that TOSS can reduce up to 86.5% of cellular traffic while guaranteeing the access delay requirements of all users. To select initial users, Barbera *et al.* established a social graph between two users if they meet each other frequently [5]. Using the social graph, they apply several well-known social structural attributes to define the social importance of a user, such as betweenness centrality and closeness centrality. Thus, the initial sources can be selected according to the social importance of users. Based on the regularity of human mobility, Li *et al.* proposed NodeRank which uses encounter characteristics to choose target nodes [38]. Encounter characteristics (including the encounter frequency, contact time and intercontact time) can, to a certain extent, ensure the integrity and availability of message delivery.

Aside from the selection of the initial users, forwarding messages is also addressed. In order to tackle this problem in a large-scale scenario, Lu *et al.* presented an offloading framework called Distance-based Opportunistic Publish/Subscribe (DOPS) which exploits distance information to make a decision on whether to start the offloading process [41]. For example, if content is associated with a three-day delay threshold, the distance the provider has traveled in the last three days is extracted. Another distance-based offloading scheme was proposed by Sun-Hyun Kim *et al.*, who used Received Signal Strength Indicator (RSSI) to roughly estimate the location of the message destination [33]. RSSI can be transferred to the distance between a user and the nearest base station.

There have been numerous papers on mobile social networks. SimBet adopts the betweenness and similarity for data forward decision [22]. Compared to other centrality calculation methods, betweenness is better for controlling the spread of the message. SimBetAge introduces an aging factor [40] which redefines the similarity and introduces flow betweenness, directed betweenness and other metrics, to deal with the dynamics of social networks and adjust social network metrics. BUBBLE borrows the concept of distributed communities to bubble messages up to the target community [30]. Literature [61] has introduced four social-aware data diffusion schemes according to the data similarity of the contacts' social relationship. Fruitful works in this field can be referred to for more details [10].

Most of the previous works concentrated on content downloading scenarios. In general, the amount of traffic generated during the down-loading process is much larger than that in the uploading process. With the popularity of social software and cloud storage technology, people are uploading photos and videos more frequently than before. Therefore, the amount of uploaded traffic cannot be ignored anymore. RoCNet exploits the opportunistic network for the uploading scenario presented in [31]. Each mobile user maintains a value called coefficient variation (CV) to record the complexity of the user's mobility. A user frequently shuttling back and forth between high-load and low-load areas will be attached to a larger CV. It is shown that the RoCNet can divert about 20% of traffic from high-load to low-load areas during peak time. This system actually operates by transferring data from the congested cell to another non-congested cell, instead of reducing the total cellular data.

4   Mobile Data Offloading Metrics

To tackle different offloading problems, existing literature proposes a variety of strategies to evaluate the offloading effect. The improvement on network performance brought by implementing offloading schemes can be reflected in a few aspects. Although both mobile users and operators benefit from mobile data offloading, they focus on two essentially diverse views of offloading metrics. In this section, we briefly discuss the performance metrics of various offloading strategies.

From the perspective of operators, mobile data offloading is expected to reduce traffic load and ease cellular congestion. While mobile users are more concerned about energy consumption and delivery delay. Here, we introduce five major evaluated metrics frequently performed as indicators of various offloading scenarios in the literature.

Offloading Efficiency: There are a few previous works using the total offloading traffic amount to evaluate the efficiency of the proposed algorithms and frameworks. However, they lack stringency in efficiency because different frameworks may generate traffic in different amounts. This is a very normal phenomenon. Given that femtocell capacity is limited, when the total traffic exceeds its capacity, more offloading is needed. Therefore, offloading efficiency is universally considered as the evaluation criterion for any offloading scheme. It is defined as the ratio of the total amount of offloaded traffic to the total amount of generated traffic without implementing offloading. It is obvious that a higher offloading efficiency is desirable.

Traffic Load Over Cellular Network: The goal of mobile data offloading technology is to relieve congestion over the cellular network. Hence, traffic load over cellular networks is another vital performance metric for operators. Due to limited capacity, each base station can only support a certain number of users. Excessive users accessing the same base station will lead to overload, which mainly reflects in a low throughput or even no available bandwidth for mobile users. An efficient offloading solution would be able to reduce the traffic load over the cellular network, especially during peak time.

Network Overhead: This metric chiefly refers to the consumption of the network resource. Taking the DTN-based offloading as an example, the flooding routing means that the source user will forward popular content to every peer user he encounters, regardless of their true requirement. Thus, these peer users may only perform as a relay. There seldom exists a direct path between the source user and the destination user, so multi-hops routing is inevitable. The storage and energy of the relay users device can be regarded as an additional occupation of network resource.

Energy Consumption: From the perspective of mobile users, battery energy consumption is a significant issue in offloading. At present, intelligent terminals are generally equipped with multiple interfaces. Searching for other devices or APs will consume part of the battery energy. However, due to the high bit rate of Wi-Fi and bluetooth, the energy consumed in the offloading process will be less than that consumed in a congested cellular network. Consequently, an energy-saving offloading mechanism is more attractive to mobile users.

Delivery Delay: Although users can accept the existence of delivery delay, an infinite delay is not ideal. As previously discussed, content is always associated with a deadline. Such a deadline represents the user's greatest delay tolerance. In most circumstances, the earlier the users receive their content within the delay-tolerance threshold, the better the offloading mechanism is. If users fail to receive content before the deadline, they will request rest data over the cellular network again. That is to say, the long waiting time is spent meaninglessly, severely harming user's satisfaction.

5   Challenges in Mobile Data Offloading

In this section, we present a discussion on the crucial challenges in mobile data offloading. Although implementing offloading strategies can bring benefits to operators, the offloading option will not be extensively accepted by mobile users, unless the potential challenges are addressed appropriately. No matter which offloading method a user chooses (Wi-Fi, femtocell or opportunistic communication), a vital concern is always the user's experience.

5.1   *Challenges in Wi-Fi offloading*

A lot of users are reluctant to spend too much time on managing how or when to start an offloading process. Thus, real-time guidance provided by operators is strongly advised in terms of user experience. A seamless service which reflects in the network handover is also important. Any disruption occurring in the forwarding or receiving process will result in strong dissatisfaction for users. For this reason, it is essential to achieve a tighter integration within 3GPP and wireless networks [53]. The Wi-Fi offloading solution is able to transfer data by bypassing the cellular network. However, it is absolutely impossible that all of the Wi-Fi APs in a city are installed by operators. There are a considerable number of APs which are completely separated from the core network, which may result in a lack of operator's supervision. Moreover, outdoor Wi-Fi APs are not widely available in most countries. Therefore, there is still a long way to go for operators who wish to implement offloading solutions via Wi-Fi networks on a large scale. Even if this dream came true in the near future, effectively managing the roaming among numerous Wi-Fi APs would be another challenge to address.

5.2 *Challenges in femtocell offloading*

Due to the fact that femtocells operate on the same frequency as the macrocellular network, the foremost challenge in femtocell offloading solutions is interference management. Here, interference includes two types: interference between macrocells and femtocells as well as inter-femtocell interference [60]. Different from the macronetwork, femtocells are often installed without a considerate plan, thereby leading to an imbalance in development density. In a dense development, a femtocell creates severe mutual interference to nearby femtocells. Similarly, a femtocell installed in a residence will create downlink interference to a device that is close but not connected to it. For more details on downlink and uplink interference, interested readers can refer to [60].

5.3 *Challenges in opportunistic communication offloading*

Privacy and security are the main challenges in opportunistic communication offloading solutions. A mobile user will never accept a strange peer to arbitrarily search and obtain the content cached in his terminal without permission. To this end, future researchers could pay more attention to privacy and security protection. Operators could design an application for users to install on their handsets. This application is responsible for identifying the content that the user is willing to share with others so that other users will not be able to search the content without a special identification. The flooding-based schemes are not viable in real life. In fact, not everyone is delighted to participate in the offloading process as a relay node, regardless of the energy and storage consumption. Therefore, it is necessary for operators to propose incentive strategies and reward the volunteer in the offloading process. Making participation more attractive can encourage users to realize the potential of offloading. Related studies are available in [55, 67].

6 Conclusion

Mobile data offloading is regarded as the natural alternative to alleviate congestion on the cellular network. Not only can operators benefit from this emerging technology through low monetary cost but users are also able to obtain a better experience. In this chapter, we posed a brief summary on the state-of-the-art offloading mechanisms. The two main classifications we have presented: infrastructure-based and infrastructureless, can almost cover all emerging use cases-driven offloading systems. For each category, detailed discussion on offloading solutions could provide enlightenment for beginners in this field. Simultaneously, we analyzed the performance metrics exploited to evaluate the offloading system. The potential challenges indicate the key issues to be addressed in the community.

Acknowledgments

This chapter is supported by the National Natural Science Foundation (61102105), the National Research Foundation for the Doctoral Program of Higher Education of China (20102304120014), the Natural Science Foundation of Heilongjiang Province of China (F201029) and the Fundamental Research Funds for the Central Universities (HEUCF1408).

References

[1] Ahn, C.-W., & Chung, S.-H. (2017). SDN-based mobile data offloading scheme using a femtocell and wifi networks. *Mobile Information Systems*, 2017, 1–15.

[2] Aijaz, A., Aghvami, H., & Amani, M. (2013). A survey on mobile data offload- ing: technical and business perspectives. *IEEE Wireless Communications*, 20(2), 104–112.

[3] Asadi, A., & Mancuso, V. (2013). Wifi direct and lte d2d in action. In *Wireless Days (WD)*, Vol. 143. IEEE, pp. 1–8.

[4] Balasubramanian, A., Mahajan, R., & Venkataramani, A. (2010). Augmenting mobile 3g using wifi. In *Proceedings of the 8th International Conference on Mobile Systems, Applications, and Services (ACM)*, San Francisco, USA, pp. 209–222.

[5] Barbera, M. V., Stefa, J., Viana, A. C., de Amorim, M. D., & Boc, M. (2011). VIP delegation: Enabling vip to offload data in wireless social mobile networks. In *IEEE Conference on Distributed Computing in Sensor Systems and Workshops (D-COSS)*, Barcelona, Spain, pp. 1–8.

[6] Bastug, E., Bennis, M., & Debbah, M. (2014). Social and spatial proactive caching for mobile data offloading. In *IEEE International Conference on Communications Workshops (ICC)*, Sydney, NSW, Australia, pp. 581–586.

[7] Boccardi, F., Heath, R. W., Lozano, A., Marzetta, T. L., & Popovski, P. (2014). Five disruptive technology directions for 5g. *IEEE Communications Magazine*, 52(2), 74–80.

[8] Burgess, J., Gallagher, B., Jensen, D., & Levine, B. N. (2006). Maxprop: Routing for vehicle-based disruption-tolerant networks. In *Proceedings IEEE INFOCOM 2006, 25TH IEEE International Conference on Computer Communications (IEEE)*, Barcelona, Spain, pp. 1–11.

[9] Calin, D., Claussen, H., & Uzunalioglu, H. (2010). On femto deployment architectures and macrocell offloading benefits in joint macro-femto deployments. *IEEE Communications Magazine*, 48(1), 26–32.

[10] Cao, Y., & Sun, Z. (2013). Routing in delay/disruption tolerant networks: A taxonomy, survey and challenges. *IEEE Communications Surveys Tutorials*, 15(2), 654–677.

[11] Cao, Y., Sun, Z., Wang, N., Yao, F., & Cruickshank, H. (2013). Converge-and-diverge: A geographic routing for delay/disruption-tolerant networks using a delegation replication approach. *IEEE Transactions on Vehicular Technology*, 62(5), 2339–2343.

[12] Chandhar, P., & Das, S. S. (2013). Analytical evaluation of offloading gain in macrocell-femtocell OFDMA networks. In *Vehicular Technology Conference (VTC Spring)*, Dresden, Germany, pp. 1–6.

[13] Chandrasekaran, G., Wang, N., & Tafazolli, R. (2015). Caching on the move: Towards d2d-based information centric networking for mobile content distribution. In *40th IEEE Conference on Local Computer Networks (LCN)*, Clearwater Beach, pp. 312–320.

[14] Chandrasekhar, V., Andrews, J. G., & Gatherer, A. (2008). Femtocell networks: A survey. *IEEE Communications Magazine*, 46(9), 59–67.

[15] Chen, Q., Yu, G., Maaref, A., Li, G. Y., & Huang, A. (2016). Rethinking mobile data offloading for lte in unlicensed spectrum. *IEEE Transactions on Wireless Communications*, 15(7), 4987–5000.

[16] Cheng, F., Yu, Y., Zhao, Z., Zhao, N., Chen, Y., & Lin, H. (2017). Power allocation for cache-aided small-cell networks with limited backhaul. *IEEE Access*, 5, 1272–1283.

[17] Cheng, N., Lu, N., Zhang, N., Shen, X. S., & Mark, J. W. (2014). Opportunistic wifi offloading in vehicular environment: A queueing analysis. In *IEEE Global Communications Conference (GLOBECOM)*, Austin, Tx, USA, pp. 211–216.

[18] Cheng, N., Lu, N., Zhang, N., Shen, X. S., & Mark, J. W. (2014). Vehicular wifi offloading: Challenges and solutions. *Vehicular Communications*, 1(1), 13–21.

[19] Cheng, N., Lu, N., Zhang, N., Zhang, X., Shen, X. S., & Mark, J. W. (2016). Opportunistic wifi offloading in vehicular environment: A game-theory approach. *IEEE Transactions on Intelligent Transportation Systems*, 17(7), 1944–1955.

[20] Cheung, M. H., & Huang, J. (2015). Dawn: Delay-aware wi-fi offloading and network selection. *IEEE Journal on Selected Areas in Communications*, 33(6), 1214–1223.

[21] Cisco, V. (2017). Cisco visual networking index: Global mobile data traffic forecast update. 2016–2021 white paper, Cisco Public Information. https://www.cisco.com/c/en/us/solutions/collateral/service-provider/visual-networking-index-vni/mobile-white-paper-c11-520862.html.

[22] Daly, E. M., & Haahr, M. (2007). Social network analysis for routing in disconnected delay-tolerant manets. In *Proceedings of the 8th ACM International Symposium on Mobile Ad hoc Networking and Computing (ACM)*, Montreal, Quebec, Canada pp. 32–40.

[23] Dimatteo, S., Hui, P., Han, B., & Li, V. O. (2011). Cellular traffic offloading through wifi networks. In *IEEE Eighth International Conference on Mobile Ad-Hoc and Sensor Systems*, Valencia, Spain, pp. 192–201.

[24] Ding, A. Y., Hui, P., Kojo, M., & Tarkoma, S. (2012). Enabling energy-aware mobile data offloading for smartphones through vertical collaboration. In *Proceedings of the 2012 ACM Conference on CoNEXT Student Workshop (ACM)*, Nice, France, pp. 27–28.

[25] Elhami, G., Zehni, M., & Pakravan, M. R. (2015). A survey on heterogeneous access networks: Mobile data offloading. In *23rd Iranian IEEE Conference on Electrical Engineering (ICEE)*, Tehran, Iran, pp. 379–384.

[26] Gonzalez, M. C., Hidalgo, C. A., & Barabasi, A.-L. (2008). Understanding individual human mobility patterns. *Nature*, 453(7196), 779–782.

[27] Ha, S., Sen, S., Joe-Wong, C., Im, Y., & Chiang, M. (2012). Tube: Time-dependent pricing for mobile data. *ACM SIGCOMM Computer Communication Review*, 42(4), 247–258.

[28] Han, B., Hui, P., Kumar, V., Marathe, M. V., Pei, G., & Srinivasan, A. (2010). Cellular traffic offloading through opportunistic communications: A case study. In *Proceedings of the 5th ACM Workshop on Challenged Networks (ACM)*, New York, pp. 31–38.

[29] Han, C., Dianati, M., Tafazolli, R., & Liu, X. (2012). A novel distributed asynchronous multichannel mac scheme for large-scale vehicular ad hoc networks. *IEEE Transactions on Vehicular Technology*, 61(7), 3125–3138.

[30] Hui, P., Crowcroft, J., & Yoneki, E. (2011). Bubble rap: Social-based forwarding in delay-tolerant networks. *IEEE Transactions on Mobile Computing*, 10(11), 1576–1589.

[31] Izumikawa, H., & Katto, J. (2013). Rocnet: Spatial mobile data offload with user-behavior prediction through delay tolerant networks. In *IEEE Wireless Communications and Networking Conference (WCNC)*, pp. 2196–2201.

[32] Kang, X., Chia, Y.-K., Sun, S., & Chong, H. F. (2014). Mobile data offloading through a third-party wifi access point: An operator's perspective. *IEEE Transactions on Wireless Communications*, 13(10), 5340–5351.

[33] Kim, S.-H., & Han, S.-J. (2012). Contour routing for peer-to-peer dtn delivery in cellular networks. In *Fourth International Conference on Communication Systems and Networks (COMSNETS 2012)*, Bangalore, India, pp. 1–9.

[34] Lee, J., Yi, Y., Chong, S., & Jin, Y. (2014). Economics of wifi offloading: Trading delay for cellular capacity. *IEEE Transactions on Wireless Communications*, 13(3), 1540–1554.

[35] Lee, K., Lee, J., Yi, Y., Rhee, I., & Chong, S. (2013). Mobile data offloading: How much can wifi deliver? *IEEE/ACM Transactions on Networking (TON)*, 21(2), 536–550.

[36] Lei, L., Zhong, Z., Lin, C., & Shen, X. (2012). Operator controlled device-to- device communications in lte-advanced networks. *IEEE Wireless Communications*, 19(3), 96–104.

[37] Li, Y., Qian, M., Jin, D., Hui, P., Wang, Z., & Chen, S. (2014). Multiple mobile data offloading through disruption tolerant networks. *IEEE Transactions on Mobile Computing*, 13(7), 1579–1596.

[38] Li, Z., Shi, Y., Chen, S., & Zhao, J. (2015). Cellular traffic offloading through opportunistic communications based on human mobility. *TIIS*, 9(3), 872–885.

[39] Lindgren, A., Doria, A., & Schelén, O. (2003). Probabilistic routing in intermittently connected networks. *ACM SIGMOBILE Mobile Computing and Communications Review*, 7(3), 19–20.

[40] Link, J. A. B., Viol, N., Goliath, A., & Wehrle, K. (2009). Simbetage: Utilizing temporal changes in social networks for delay/disconnection tolerant networking. In *Mobile and Ubiquitous Systems: 6th Annual IEEE International Conference on Networking & Services, MobiQuitous' 09*, Toronto, ON, Canada, pp. 1–2.

[41] Lu, X., Lio, P., & Hui, P. (2016). Distance-based opportunistic mobile data offloading. *Sensors*, 16(6), 878.

[42] Mayer, C. P., & Waldhorst, O. P. (2011). Offloading infrastructure using delay tolerant networks and assurance of delivery. In *2011 IFIP (IEEE) Wireless Days (WD)*, Niagara Falls, ON, Canada, pp. 1–7.

[43] Mehmeti, F., & Spyropoulos, T. (2017). Performance analysis of mobile data offloading in heterogeneous networks. *IEEE Transactions on Mobile Computing*, 16(2), 482–497.

[44] Ngo, M. H., & Krishnamurthy, V. (2009). Optimality of threshold policies for transmission scheduling in correlated fading channels. *IEEE Transactions on Communications*, 57(8), 2474–2483.

[45] Onoue, Y., Tamai, M., & Yasumoto, K. (2014). Energy-constrained wi-fi offloading method using prefetching, In *IEEE 79th Vehicular Technology Conference (VTC Spring)*, pp. 1–5.

[46] Pieterse, H., & Olivier, M. S. (2014). Bluetooth command and control channel. *Computers & Security*, 45, 75–83.

[47] Poularakis, K., Iosifidis, G., Pefkianakis, I., Tassiulas, L., & May, M. (2016). Mobile data offloading through caching in residential 802.11 wireless networks. *IEEE Transactions on Network and Service Management*, 13(1), 71–84.

[48] Qutqut, M. H., Al-Turjman, F. M., & Hassanein, H. S. (2013). MFW: Mobile femtocells utilizing wifi: A data offloading framework for cellular networks using mobile femtocells. In *IEEE International Conference on Communications (ICC)*, pp. 6427–6431.

[49] Rebecchi, F., de Amorim, M. D., & Conan, V. (2014). DRoid: Adapting to individual mobility pays off in mobile data offloading. In *IFIP (IEEE) Networking Conference*, Trondheim, Norway, pp. 1–9.

[50] Rebecchi, F., de Amorim, M. D., & Conan, V. (2016). Should i seed or should i not: On the remuneration of seeders in d2d offloading. In *IEEE 17th International Symposium on World of Wireless, Mobile and Multimedia Networks (WoWMoM)*, Coimbra, Portugal, pp. 1–9.

[51] Rebecchi, F., De Amorim, M. D., Conan, V., Passarella, A., Bruno, R., & Conti, M. (2015). Data offloading techniques in cellular networks: A survey. *IEEE Communications Surveys & Tutorials*, 17(2), 580–603.

[52] Ristanovic, N., Le Boudec, J.-Y., Chaintreau, A., & Erramilli, V. (2011). Energy efficient offloading of 3G networks. In *IEEE Eighth International Conference on Mobile Ad-Hoc and Sensor Systems*, Valencia, Spain, pp. 202–211.

[53] Sou, S.-I. (2013). Mobile data offloading with policy and charging control in 3gpp core network. *IEEE Transactions on Vehicular Technology*, 62(7), 3481–3486.

[54] Spyropoulos, T., Psounis, K., & Raghavendra, C. S. (2005). Spray and wait: An efficient routing scheme for intermittently connected mobile networks. In *Proceedings of the 2005 ACM SIGCOMM Workshop on Delay-tolerant Networking (ACM)*, pp. 252–259.

[55] Wang, K., Lau, F. C., Chen, L., & Schober, R. (2016). Pricing mobile data offloading: A distributed market framework. *IEEE Transactions on Wireless Communications*, 15(2), 913–927.

[56] Wang, T., Cao, Y., Zhou, Y., & Li, P. (2016). A survey on geographic routing protocols in delay/disruption tolerant networks (DTNS). *International Journal of Distributed Sensor Networks*, 6, 8.

[57] Wang, T., Sun, Y., Song, L., & Han, Z. (2015). Social data offloading in d2d-enhanced cellular networks by network formation games. *IEEE Transactions on Wireless Communications*, 14(12), 7004–7015.

[58] Wang, X., Chen, M., Han, Z., Wu, D. O., & Kwon, T. T. (2014). Toss: Traffic offloading by social network service-based opportunistic sharing in mobile social networks. In *Proceedings of IEEE INFOCOM 2014-IEEE Conference on Computer Communications*, Toronto, ON, pp. 2346–2354.

[59] Xiaofeng, L., Pan, H., & Lio, P. (2013). Offloading mobile data from cellular networks through peer-to-peer wifi communication: A subscribe-and-send architecture. *China Communications*, 10(6), 35–46.

[60] Yavuz, M., Meshkati, F., Nanda, S., Pokhariyal, A., Johnson, N., Raghothaman, B., & Richardson, A. (2009). Interference management and performance analysis of umts/hspa+ femtocells. *IEEE Communications Magazine*, 47(9), 102–109.

[61] Zhang, Y., Gao, W., Cao, G., Porta, T. L., Krishnamachari, B., & Iyengar, A. (2012). *Social-Aware Data Diffusion in Delay Tolerant MANETs*. Berlin: Springer, pp. 457–481, ISBN 978-1-4614-0857-4.

[62] Zhao, N., Liu, X., Yu, F. R., Li, M., & Leung, V. C. (2016). Communications, caching, and computing oriented small cell networks with interference alignment. *IEEE Communications Magazine*, 54(9), 29–35.

[63] Zhao, N., Yu, F. R., & Leung, V. C. (2015). Opportunistic communications in interference alignment networks with wireless power transfer. *IEEE Wireless Communications*, 22(1), 88–95.

[64]   Zhao, N., Yu, F. R., & Leung, V. C. (2015). Wireless energy harvesting in interference alignment networks. *IEEE Communications Magazine, 53*(6), 72–78.

[65]   Zhao, N., Zhang, X., Yu, F. R., & Leung, V. C. M. (2017). To align or not to align: Topology management in asymmetric interference networks. *IEEE Transactions on Vehicular Technology, 66*(8), 7164–7177.

[66]   Zhu, Z., Fan, Z., Cao, F., & Sun, Y. (2011). Data offloading method for interference and mobility management in femtocell systems. In *7th IEEE International Wireless Communications and Mobile Computing Conference (IWCMC)*, Istanbul, Turkey, pp. 2115–2120.

[67]   Zhuo, X., Gao, W., Cao, G., & Dai, Y. (2011). Win-coupon: An incentive framework for 3G traffic offloading. In *19th IEEE International Conference on Network Protocols*, Vancouver, BC, Canada, pp. 206–215.

Chapter 8

Exploring the Scalable Event-Driven Load Balancer Using *Hybrid Intelligence* for Smart Paradigms

Soumya Banerjee[*,§], Philipp Brune[†,¶] and Youakim Badr[‡,‖]

CNAM-CEDRIC Lab, Paris, France
†*The Neu-Ulm University of Applied Science, Germany*
‡*INSA de Lyon LIRIS Lab, France*
§*soumyabanerjee@bitmesra.ac.in*
¶*Philipp.Brune@hs-neu-ulm.de*
‖*youakim.badr@insa-lyon.fr*

1 Introduction

In the multiple and multi-objective scenario of internet of things (IoT) implementation, it is appropriate to achieve the quality of services (QoS) without affecting the distribution of the rationing of resources across various applications. In the case of back-end presentation of wireless sensor network (WSN) concerning smart tag-based applications, it is also imperative to devise certain strategic solutions. This is because the design of the conventional WSN architecture and corresponding load-balancing technology becomes more complex when addressing the requirements of adaptability and high flexibility with respect to abrupt changes [68]. In reality, any smart paradigm comprises different load-driven applications. Thus, they can demonstrate a variety of applications, e.g., smart meters, mobility of electrical services, demand service management, data distribution across data hub and renewable integration services [69]. Therefore, it is worth discussing connected objects and the distributed system to interconnect their compatibility of resource mapping. The backdrop of such paradigms has been well defined through IoT integration. However, the flavor of load balancing has always been intensified toward the mobile *ad hoc* network [70].

In the next 30 years, it is projected that there will be over 100 billion internet-connected devices that fall under the rubric of the IoT [1, 21]. These include smart phones, TVs, cars, robots and wearable computers, just to mention a few. The first direct consequence of the IoT is the deluge of data generated by every physical or virtual device. The deployment of IoT requires scalable computing capabilities to collect, store and process large volumes of data typically referred to as Big Data. One of the fundamental challenges for IoT is managing the exponentially growing number of devices and scaling up the collection of massive amounts of data (i.e., "billions of records per second") and then distributing them across multiple computing resources. In the last decade, a variety of system architectures have been implemented for data-intensive applications, such as popular websites, large internet relay chat (IRC) networks, domain name system (DNS) servers and parallel databases. These architectures mainly rely on load balancing as dedicated scheduling algorithms to divide incoming data traffic over clusters of back-end servers. Several scheduling algorithms have been proposed for load balancing (e.g., join-the-shortest-queue, min–max, max–min, fastest processor to largest task first (FPLTF) and round robin). More sophisticated algorithms have also been developed to take additional constraints (e.g., least response times, number of connections, single or parallel servers) and optimize some objective functions as "when" and to "which" server the collected data should be to sent to [22, 23]. The load balancing evolves within the context of cloud computing and leads to *MapReduce* frameworks like Hadoop as the popular way to handle scalable and massive datasets. Advanced research in cloud computing has focused on streams of well-structured data with multiple constraints (e.g., resources loads, network switches and storage, number of clusters and cluster size) and dynamic load balancing to be performed periodically during a runtime in ReduceMap [24, 25]. Regardless of the fact that load balancing remains a non-polynomial (NP) hard problem, current load balancers are not applicable to the IoT [26]. First, they are mainly based on heuristic rules and a predefined set of constraints and parameters, excluding possibilities to add new constraints or deal with the missing values for their parameters at runtime. Second, they are not designed to manage huge volumes of unstructured data from different devices. In large-scale environments, multiple reasons drastically limit the scalability and reliability of current load balancers. This includes not being aware of the heterogeneity of the data sources and types, which might impact schedule decisions at runtime. This possibility is based on the assumption that (1) constraints on the software and hardware are always measurable and (2) neglecting the random behavior and uncontrollable alterations (e.g., security threats, degradation mode, hardware and software failures), or the process cannot ensure optimal load balancing when resources dynamically evolve (e.g., cloud resources elasticity) or dynamically configure resources to maintain optimal load balancing. In addition, the IoT relies on machine-to-machine (M2M) technologies to ensure communications across connected devices and develop a wide range of applications, such as industrial automation, telemetry, logistics, smart grid and smart cities, mostly for monitoring and also for control purposes. Nevertheless, existing M2M solutions are highly fragmented and miss interoperability, because of which the load-balancing encounters technical barriers. These drawbacks and limitations urge the development of a new class of load balancers for the IoT, making a shift from a one-size-fits-all load-balancing strategy for servers and databases to on-the-fly load-balancing strategies adapted to M2M infrastructures. In response to these challenges, we propose a dynamic load-balancing infrastructure based on the international M2M standard (a.k.a one M2M). This is presented in many recent researches with respect to the long-range wide-area network (LoRaWAN) standard [66]. Therefore, an adaptive load balancer must be aware of the unpredictable changes in devices and should be compatible to the evolution of large-scale environments. Considering the postulates of IoT contexts, the concept of hybrid intelligence is introduced with respect to optimization and behavior of social insects. The relevance of the incepting hybrid intelligence is bi-focal: first, the optimization of allocation of positioning of a particular physical machine to virtual machine (VM) could be achieved, which in turn could reduce the computational cost. Subsequently, the leader-and-follower strategy of social insects may introspect appropriate synergy of near-optimal mapping from a physical machine to a VM under any distributed environment. The proposed concept provides an effective algorithm to trace the strategy for physical to virtual load balancing, and finally, the dynamic load-balancing technique is compared with certain benchmark test suits of load-balancing and mapping heuristics. The remaining part of the work has been organized as follows. Section 2 discusses similar works concerning the incorporation of computational intelligence and optimization in load-balancing methodologies, followed by the gist of the role of hybrid intelligence in the present model in Section 3.

Sections 3.1 and 3.1.1 describe the event model and core of service-oriented load balancer architecture. Section 3.2 elaborates the context of dynamic load balancing based on multi-objective optimization. A brief referential backdrop has been given in Section 4 on bi-level optimization and bio-inspired techniques. Section 4.1 details the notions, semantics and pseudocode of the proposed algorithm followed by the relevance of the experimental dataset and the acceptable attribute to simulate the iterations in Section 4.2. Section 5 discusses the results and observations accomplished through the procedure. Importantly, this section refers the comparison of results with the benchmark allocation heuristics in the IoT or M2M paradigm. Finally, Section 6 summarizes contributions and mentions the future scope of intelligent load balancers in the case of an event-driven environment.

2 Related Works

With the increasing adoption of cloud computing and cloud services, ensuring the QoS of these services has become a major challenge for cloud providers. While machine learning (ML) techniques may be applied to predict and detect the upcoming QoS outages [65], to actually ensure the required response times and scalability of their services, appropriate and adaptive resource scheduling and load balancing are the core requirements for every cloud data center [55, 58]. Ideally, a load balancer can thus be defined as a device that acts as a reverse proxy and navigates the load traffic across a number of servers [4]. Load balancers are used to increase the capacity (concurrent users) and reliability of applications. In the context of streaming cloud load balancing, where data have to be processed by continuous queries and results made available on the fly, it is worth customizing the cloud system with certain heuristic and computationally intelligent backups [4]. Usually, the load is distributed in a cloud data center by mapping resources from the physical machine to the VMs [41]. ML techniques, conventionally in the form of classification, are used to create groups of VMs based on their CPU and RAM utilization. They also quantify user jobs/tasks into different groups based on their size and information from log files. This is also a suitable basis to deploy computational intelligence for necessary classification [43]. In present days' interconnected environment, for the IoT, the mapping and managing of dynamic structural load solicits different components of intelligence and belief. Deep Belief Network is one of them. Practically, they can be used for modeling different interconnected objects and entities. The deployed load from extra layers can acquire features from the low-ranked layers. Finally, such a network load is balanced and distributed across the interconnected objects [44]. Research efforts are reflected in the form of parallel programming and cloud-driven load distribution [45, 46]. In addition to load balancing, load forecasting is another research application where computational intelligence can be used [47, 65]. The mobility load-balancing (MLB) concept can reconfigure the switchover parameters to consolidate traffic load depending upon resource deployment. Conventional ML algorithm, e.g., support vector machine and Monte Carlo simulation, can predict radio resource utilization [48]. In recent years, the topic of load balancing for cloud computing has been intensively studied in the literature. Since load balancing in general is an optimization problem, various optimization algorithms have already been applied and evaluated [55]. Multiple recent metastudies and literature reviews also compare and analyze existing approaches [5, 23, 25, 53, 55, 56, 58, 61]. Most known scheduling and optimization techniques have been applied to cloud load balancing over the years [53, 55], ranging from classical scheduling algorithms — such as join-the-shortest-queue, min–max, max–min, or round robin — to fuzzy logic or genetic algorithms (GAs) [10, 11, 17, 20, 28, 31, 35, 36, 39, 40, 52, 54, 63]. Bio-inspired algorithms have also been applied with some success to the load-balancing problem, e.g., ant-colony, interaction artificial bee colony, firefly or particle-swarm optimization [3, 6–9, 13, 29, 36, 59, 60, 62, 64]. The dynamic nature of bio-inspired algorithms is beneficial for balancing the load between different variable instances. This is visible in continuous financial data stems and layouts, where the arrival of different combinatorial parameters are not following a steady pattern during the whole day of trading [2]. The workload seldom attains multiple quanta of the average load. Traditional algorithms of optimization like GAs could not be the best fit as they do not explore the local minima issues while searching the space of load [10]. Hence, bio-inspired algorithms can be a worthy choice to formulate an effective load scheduler. While many proposed applications try to optimize the load with respect to a single objective (e.g., average load per cluster node or response time), recently, multi-objective optimization has also come into focus [57]. The real challenge of load balancing is to keep the scope of multi-objective and multimodal optimization alive while satisfying the constraints [37]. Another research direction is the use of hybrid optimization strategies, combining multiple (e.g., two) optimization algorithms, e.g., classical scheduling with bio-inspired algorithm. So far, few results have been published on this topic [4, 64]. Hybridization is mainly achieved to take care of the varieties and uncertain load conditions across the VM network [11]. In modern business cloud applications like Big Data processing in the IoT context, high adaptability and cost effectivity of the cloud infrastructure are important criteria. Gufler *et al.* [15] proposed the model and the requirement for a load-balancing component in MapReduce. The data distribution is random. They coined Top Cluster, a distributed monitoring system for investigating data skewness in MapReduce systems. Improvized through distributed top-k algorithms, Top Cluster is tuned to map with the characteristics of MapReduce frameworks [15]. Similarly, Hadoop-driven Flubber has also been proposed with respect to four adaptive load-balancing techniques [16]. Bhattacharya *et al.* discussed the evolutionary approach in the context of Big Data and IoT [18]. Buyya *et al.* [19] discussed the utility of the cloud simulation toolkit. While more advanced approaches such as multi-objective or hybrid algorithms have been discussed for the load-balancing problem, their combination to implement a more advanced, adaptive and flexible load-balancing scheme in a generic way, e.g., being capable of addressing the needs of highly distributed IoT environments, still remains an open issue. Therefore, in this work, the following question is addressed: How might such an approach be realized by proposing and evaluating a novel hybrid, bi-level, bio-inspired multi-objective load balancing algorithm [42]?

3   Role of Hybrid Intelligence

Unlike traditional and cloud load-balancing approaches, the proposed dynamic load-balancing infrastructure seeks to develop scalable load balancing strategies for the IoT with an awareness of the contextual changes under uncertainty and self-adaptation capabilities. It advocates a multidisciplinary solution through a unified approach, integrating an event model to specify a rich data structure to describe the collected data from the connected devices. Services to encapsulate M2M functionalities and compose them dynamically to deploy a balancing strategy multi-objective optimization algorithm are always treated as technical challenges. However, without these compositions, it will be difficult to build dynamic load-balancing strategies for distributing on event payload or on distributed Hadoop storage contexts. The mentioned dynamic load-balancing infrastructure thus relies on three different yet complementary research topics, soft computing, service computing and ML in the context of IoT and massive real-time data. Scalability is ensured by continuously adapting the load balancing by means of learning techniques in response to changes due to the heterogeneity of event content, dynamics of the context and elasticity of the computational resources. The adaptation relies on soft computing to build optimal scheduling strategies by approximate reasoning with fuzziness and uncertainty and on service computing to dynamically reconfigure computational resources to deploy these strategies at runtime. The role of hybrid intelligence is evident from the uncertain and variable load conditions. The trend of hybrid algorithm could be a simple load-balancing algorithm used by the distributive node to distribute the received tasks efficiently over the VM under a normal work load by finding the optimal VM among the group of VMs; it will assign the load in a heterogeneous distributed computing environment. Considering the magnitude of uncertain contexts for the IoT paradigm [37], it becomes a practice that the hybrid algorithm consists of both random and greedy components. We investigate the most frequently used load balancer algorithms in implementation.

Load-balancing solutions can be divided into hardware-based load balancers and software-based load balancers [40]. Hardware-based load balancers are compatible with the high-speed network traffic, including application-specific integrated circuits (ASICs) toward specific application. Software-based load balancers are reciprocative with conventional operating system and hardware.

Data center controller assigns the requests over the list of VMs through the distributed time segments. However, the allocation of request is circular in nature. After the very first request to VM is relinquished, the next priority of request could be released. Definitely, an appropriate indexing of each state, either free or occupied, is maintained in VMs. The objective of the algorithm is to find a specific VM to receive a request for further processing.

3.1 *Event-driven model*

For a diversified load and a multiple server, there are dynamic file grouping and event-driven components of distributed computing. Therefore, it is essential to assist the storage node and I/O bandwidth by using dynamic grouping events [49].

Load balancing under intra-cluster availability can also be a challenging problem due to diversified cache memory and bandwidth to handle a particular request for such available media servers. Resource optimization can also become an important task for the cluster with limited network bandwidth, cache capacity and disk I/O bandwidth, without sacrificing the performance index.

3.1.1 *Service-oriented architecture (SOA) in a load balancer*

In web service discovery, there are certain default standards to facilitate web service and communication protocols. Universal Description, Discovery and Integration (UDDI) and Simple Object Access Protocol (SOAP) are some of these protocols [67].

Service-oriented architecture (SOA) is an orchestrated architecture in which software components are formulated as reusable services, which are flexible and can be utilized with a notable potential of platform independence features. Therefore, SOA is a set of high-level software architecture guidelines. However, it is not the direct implication of load balancing in a distributed system. Conventionally, with SOA, the service consumer may look up a UDDI registry to locate the service provider. Some of the latest UDDI registry software provide simple (e.g., round robin) load balancing. Another SOAP idea is using WS-addressing, but it is not really meant for load balancing. Hence, the role of SOA should come up with the fault-tolerant component of a dedicated load balancer in the scenario of specific grid-computing [50, 51].

3.2 *Dynamic load balancing based on multi-objective optimization*

For dynamic load balancing, in the context of the IoT and Big Data, the concept of self-organization has been deployed. The concept also helps to formulate the strategy of the hybrid intelligent technique. In this paradigm, the proposed model refers the seminal work of Bonabeau *et al.*, which has been modeled with respect to the given problem and yields the following:

(a) **Positive Feedback:** To maintain a convenient local load distribution structure across the system, a portion of information is retrieved in the form of feedback mechanism. It supports diversified load distribution conditions.

(b) **Negative Feedback:** It is the nullification effect to be conceived by the positive feedback strategy under load condition.

(c) **Fluctuation of Random Load Distribution:** This could be a possible measure to avoid system/event stagnation.

(d) **Multiple and Multilevel Interaction:** The proposed hybrid intelligent system is capable of learning patterns from different load characteristics.

Social behaviors of different insects [38] improvize the principle of a self-organization of a dynamic event and elasticity of the computational resources. Several interesting propositions are identified while establishing an ideal blend of bi-level optimization and bio-inspired techniques. The model considers principally different phases defining the leader-and-follower concept in the existing load distribution scenario.

4 Background of Bi-level Optimization and Bio-inspired Technique

The background of choosing the blend of bi-level optimization and bio-inspired techniques is referred to describe their corresponding relevance.

The distinguishing factor between bi-level programming problems and other optimization problems is their structural configuration. A bi-level program is also an optimization problem, which comprises other optimization processes leading with the constraints as an external layer of optimization. The external optimization problem is the lead-level optimization task, whereas the constraining optimization task has been considered as the follower-level problem. This embedded structure implies that a solution to the leader problem can only be acceptable, if there persists an optimal solution to the lower-level problem too. Mathematically, the expression can be given as follows:

$$\sum{}_X \operatorname{Max} F(X, Y) \quad \text{subject to } G(X, Y) \le 0,$$

$$\text{and } \operatorname{Max} \sum{}_Y f(X, Y) \quad \text{subject to } g(X, Y) \le 0,$$

where $F, f : R^{n_1} \times R^{n_2} \to R^P$ and $g : R^{n_2} \to R^q$ (constraints for leader and follower) [42] and $X \in R^{n_1}$ & $Y \in R^{n_2}$ (n_1 and n_2 are the dimensional decision variables).

Initially, the proposed hybrid algorithm yields an optimal yet homogeneously distributed initial population of N number of bio-inspired agents, where each bio-inspired agent $\text{BA}_i(i = 1, 2, \ldots, N)$ is a G-dimensional vector.

Here, G is the number of variables in the optimization problem and they can be initialized as follows:

$$\mathrm{BA}_{ij} = \mathrm{BA}_{\min j} + u(0,1) \times (\mathrm{BA}_{\max j} - \mathrm{BA}_{\min y}), \tag{8.1}$$

where $\mathrm{BA}_{\min j}$ and $\mathrm{BA}_{\max j}$ are bounds of BA_i in jth direction and $u(0,1)$ is an uniformly distributed random number in the range of $[0,1]$.

In the second level, each agent A_i modifies its current numerical value based on the load distribution of the local position. The effective fitness value is calculated by the proposed model. The condition is also proposed to check whether the system is under a balanced load or not. If the position value of the placed load in the system is higher than that of the previous one, then the proposed model updates its current position in terms of bio-inspired agents. The referential updated position for the ith agent BA_i where A is also a member of other local sub-distribution networks with respect to different variables of load can be expressed as follows:

$$\mathrm{BA}_{\mathrm{new}ij} = \mathrm{BA}_{ij} + u(0,1) \times (L\mathrm{Load}_{kj} - \mathrm{BA}_{ij}) + u(-1,1)$$
$$\times (L\mathrm{Load}_{rj} - G\mathrm{Load}_{ij}). \tag{8.2}$$

which represents the jth dimension of the rth k local subgroup with a random load under the lead load position.

BA_{rj} is the jth dimension of the rth BA_i randomly positioned kth sub-network or subgroup of load distributions, such that $r \neq i, u(0,1)$ is a random number to be distributed across the load distribution of the VM.

In this phase, the positions of load controlled leader (BA_{ij}) are augmented based on probabilities prob_i, which are calculated using their feasibility value. Therefore, a more feasible candidate may lead toward better solution. The probability prob_i is calculated using the following expression (it is a function of feasibility):

$$\mathrm{prob}_i = 0.9 \times \frac{\mathrm{Appr.Bal.}F}{\mathrm{Max.Bal.}F} + 0.1. \tag{8.3}$$

Given the resource requirements of VMs, an algorithm is expected to come up with a placement plan, which gives a mapping for a VM to be placed to a specific physical machine. Here, each x_{ij} indicates whether VM i is placed on the physical machine j. Hence, x_{ij} will be 1, if VM i is placed on placed machine (PM) j, 0 otherwise. The strategy of formulating the allocation of optimal map from physical to virtual is an optimization problem comprising several constraints, like the allocation capacity of a processor and assurance of finding the optimal map. If each physical machine of dimension j is considered, then the sum of the total required resources of all deployed VMs should be less than or equal to the total available capacity, denoted as follows:

$$\sum_i F_{ik} M_{ij} \leq 1 \qquad (8.4)$$

for all positive values of k and j, where F_{ik} is the part of the fractional resource requirement of an individual VM i along the dimension k.

It is expected that as the best case strategy, all VMs should be placed optimally. Hence, for each VM i, exactly one of the M_{ij}'s is expected to be 1, and others 0. Hence, the following expression persists:

$$\sum_i M_{ij} = 1 \qquad (8.5)$$

for all positive value of i.

Therefore, to formulate the objective function, we associate a variable y_j for each PM j as shown. y_j will be either 0 or 1 depending on whether PM j has to be used or not. It implies that $y_{ij} \geq M_{ij}$ and finally the value of y_j should be minimized.

In the basic model, we assume the total number of nodes is N and the total number of processes running in the system is m. For the purpose of illustration, both N and m remain unchanged during the execution. Let g_I denote a grid node i, where $1 \leq i \leq n$. The self-organization principle signifies the load-balancing strategy across the total number of distributed paradigms. The "paradigm" in this case consists of N agents. All are identified by their time-dependent vector of features $p_i(t)$. The time vector also relates the number of processes and the concerned work load of the processes. Each agent modifies and regulates its combined attributes by a weighted average of attributes from the adjoining agents $p_j(t)$. These weighted averages are expressed in the form of modulus of weights (essentially of positive values), $a_{ij} \geq 0$. The expression is given as follows:

$$p_i(t + \triangle t) = p_i(t) + a\triangle t \sum_{j=1}^{N} a_{ij}(p_j - p_i)[(x_i(t)), v_i(t))]. \qquad (8.6)$$

Here, agent $p_i(t)$ aligns its features at a rate proportional to a weighted average of the differences relative to its "neighbors" $\{p_j - p_i\}$. The positive parameter quantifies the frequency of alignment; addition of $x_i(t), v_i(t)$ denotes the collective flocking attribute, which is modified for every iteration as a fresh map from the physical machine to VM distribution. For different users' requests and time of service for a given load, the cloud center must maintain the QoS for varying loads [32, 40]. Performance enhancement for different types of cloud segments requires manifold load-balancing techniques. Load distribution depends on the end users' load taxonomy. Here, as per the corresponding distance from the user and respective load, the criteria of optimization are deployed. Operating systems, such as Microsoft and Linux, maintain embedded plug-ins to combat the fluctuation of load across the network. However, near-optimal load balancing has already been introduced in the form of pushing load-balancing functionality into the soft network edge (e.g., virtual switches). The soft network offers almost no changes at the transport layer or network hardware. Thus, a proximal unit of fine balancing of load can be achieved dynamically with a minimum number of corresponding changes [30].

4.1 *Data and parametric representation: Experimental settings*

To demonstrate the effective attribute of the proposed algorithm (see Table 1) in the context of dynamic load distribution, the following different load mapping and time stamps have been retrieved.[1]

Additionally, certain live data streams, e.g., http://www.w3.org/TR/hcls-dataset/ are considered.

The population size of the bio-inspired agents (BA_i), rate of balanced load distribution, global random load limit (GRLL), global and local random load limit (LRLL) and maximum number of groups (MG) to be formed are crucial parameters for the proposed model. All these parameters

Table 1. Different parameters for proposed algorithm.

Notation	Semantics
N	Number of agents
Capacity (c_i)	Capacity of system allocation in seconds
No. of processes (m_i)	Processes from physical machine to VMs
Process workload	Specified load
Bandwidth	Distribution

---

[1] https://www.npmjs.com/package/loadtest.

need to be analyzed sensitively. The tuned exact value for the scalable load-balancing system is required and hence the process size of the BA_i is varied from 50 to 190 units with a step size of 20. The rate of balanced load distribution is varied from 0.1 to 0.9 units with a step size of 0.1. Maximum group formation under the load is varied from 1 to 6 with a step size of 1. LRLL is varied from 100 to 2500 units with a step size of 200 whereas GRLL is kept within the range from 10 to 200 units with a step size of 30. The evaluation process of the final load is subject to the following assumptions:

Algorithm 1 High-level Description of the Proposed Algorithm

/*This algorithm is executed on each node g_i.*/

Algorithm (Parameter list extended)

1. Initialize the *N number of bio-inspired agent populations, LRLL, GRLL, rate of balance load distribution.*
2. Estimate fitness, i.e., the minimum distance of tasks placed in VMs from the service queue of physical machines.
3. Poll the global leader in random local distribution through greedy mode of selection following equations (8.1) and (8.2):
      while (termination criteria is not satisfied)
          do
(i) For finding the objective (distances of tasks from the service queue), self-organizes new position of tasks with respect to the variable load position using self-organized principle using equation (8.6).
(ii) Apply the most *deserving* or *hungry* event between existing position of load and newly allocated capacity c_i.
(iii) Calculate the probability by using equation (8.3)
(iv) If any local sub-group and its variable local is not updated after specified number of iterations.
(v) Then load of that group is requested according to equation (8.4):
      end if
         end while
           end for
4. Find VM with minimum service distance and maximum score.
5. Place that VM over host and remove it from VM list.
6. Update host's current capacity and present the fresh load lists.
7.   **end**

- The rate of a balanced load is verified from 0.1 to 0.9 while the value of MG, LRLL and GRLL and the size of population are fixed to 7,1550, 65 and 50, respectively.
- MG is also varied from 1 to 6 units, while LRLL, GRLL and size could be kept as their previous values.
- Rate of balanced load is increased from 0.1 to 0.4 through several iterations.
- The stopping criteria is either the maximum number of function evaluations (which is set to be 2.0×105) reached or the corresponding acceptable error.

4.2   *Benchmark comparison of the proposed load balancer algorithm*

We also observed that to tune the parameters under a distributed environment, creation of multiple looping and iterations of server is required. Practical approach suggests a specialized implementation environment for keeping the near-optimal control, while presenting the mapping from physical machine to VMs. We introspect the pytest–xdist plugin, which extends pytest with some unique test execution modes capable of delivering load-balancing structures. The strategy comprises the following:

- **Loop-on-Fail:** The selected test instances of load balancing could be executed repeatedly in a subprocess. After respective iterations, pytest waits until the final alterations are done in the file system and then repeats the predecessor test till all the critical points are addressed. After exhaustive iterations, again a complete loop execution is performed.
- **Multiprocess Load Balancing:** If multiple CPUs or hosts are available, then a combined test run should also be considered. This solicits to boost the speed and resources' deployment of remote machines.

Subsequently, the following are regarded as the criteria to measure the efficiency of strategies directed for load balancing:

- Throughput
- Overhead
- Fault tolerance
- Migration time
- Response time
- Resource utilization
- Scalability

Section 5.1 presents a comprehensive comparison with respect to the other existing algorithms of similar capacities and observed certain benchmarked components. In the usual load-balancing techniques, it becomes important to detect the different access points based on the thresholds. The threshold is assigned by either the data center or the clients [41]. Higher values with respect to resource usage could decide the different access points. However, the upper or lower threshold may imply over-utilization and or under-utilization for the load, irrespective of the dimension of resources. These conditions are being considered in the proposed consolidation and load-balancing algorithms. Practically, the findings of the proposed load balancer algorithm (principally configured with principle of self organization) are similar to those of two other existing counterparts: the first one is *Sandpiper* [33], which aims at hotspot mitigation alone and not server consolidation. Since this heuristic does not take consolidation into account, we are unsure of its performance when compared with other consolidation algorithms. Major consolidation algorithms do hotspot mitigation, which is Sandpiper's sole objective. This algorithm aims at hotspot mitigation to prevent service-level agreement violations. The upper-value access points are detected, when the CPU usage values are violated with respect to the threshold assigned to the CPU. Based on the thresholds set, PMs are distinguished as underloaded or overloaded. The strategy of sorting PMs is their volume metric itself.

The other algorithm of a similar function is *Entropy* [34], which uses a *Constraint Satisfaction Programming*-based technique to figure out the minimum possible PMs needed to house the VMs. Subsequently, it does VM placement by minimizing migration costs in the process. This algorithm is heuristic-prone in placing the VMs and can be analytically considered with the proposed load balancer technique.

All the observations and evaluations are done over a fixed time window T. During this T time frame, we log all the details while the algorithm is running and predefined workloads are generated from different clients at the VMs before the logging begins. In our work, an algorithm (say A) is defined to be better than the other (say B) for a particular scenario as given in the following:

- The time taken within that time window T, to mitigate hotspots or cold-spots or reduce the number of PMs, is lesser in A than B.
- If the incurred cost of migration in A is lesser than in B, then A is better than B, else vice versa. We have used the *Migration Cost*

Metric as the number of migrations. In the research literature, if too many frequent migrations make the system unstable, if lesser number of migrations are triggered by A than B to consolidate (and thus mitigate hotspots/coldspots), then A is better than B. Also, the more the number of migrations, the more the migration overhead. Count the number of migrations triggered by A within T time units = A count. Count the number of migrations triggered by B within T time units = B count. If A count < B count, A is better than B, else vice versa.

- If an algorithm can reduce a greater number of PMs, it is substantially better than the other in consolidating.

5 Results and Discussion

After implementing the proposed methodology and synthetic dataset as presented in the loop structure, the following results were obtained. Table 2 snaps the paradigm of load balancing of a number of tasks with prior and posterior load balance status. The number of tasks per process is fixed for the simulation. Subsequently, the strategic types of load distribution are given in Table 3. The strategies are heuristics to place the VMs over the group of physical machines under a variable load condition.

Table 2. Time span layout of prior and posterior load-balancing status.

No. of tasks	Prior load balance status (s)	Posterior load balance status (s)
10	12.6	6.9
20	28.7	16.7
30	37.2	20.8
40	58	29.2

Table 3. Load mapping from physical to virtual map.

Load distribution type (LD)	1	2	3	4	5	6	7	8	9	10	Av.	Wrap val
Dissimilar	40	41	37	36	38	41	40	41	38	40	39.2	0.667
Critical	41	41	38	37	34	37	38	39	37	39	38.1	0.663
Multi-dimensional	39	40	34	35	37	35	38	38	38	37	37.1	0.62
Multi-level	40	40	34	35	36	37	38	37	37	35	36.9	0.63
Density-based	39	40	33	35	36	35	37	37	38	36	36.6	0.61
Dot product-driven	40	40	34	35	36	37	38	37	37	35	36.9	0.61
Best optimal	55	55	55	55	55	55	55	55	55	55	55	1

The types are subjected to dissimilar or variable load, critical, multidimensional, multilevel, density-based, "dot" product-driven and optimal load distribution. Among them, density-based type describes the visits of the physical machines in the decreasing order of certain predefined functions of capacity used along all dimensions. We have implemented this function as the product of the capacity usages because multiplication is nearer to the real-world concept of "density". Similarly, the requirements of VMs along the specified dimensions and physical machine capacity usages along those dimensions are expressed as vectors. Then, a dot product of these two vectors is taken as heuristics for placing toward each physical machine. The physical machines are visited in decreasing order of this dot product. The proposed best optimal is given in the last column in Table 3. Wrap value is the indicator ratio of the average number of physical machines used by the algorithm to the average number of physical machines used by the scalable load balancer (Fig. 1).

As can be seen from the graph in Fig. 2, when more heavier VMs got included in the VM list, the number of physical machines used by the heuristics increased. In output vs. time of execution, different load strategies are addressed. The behavior of load can be abruptly interrupted with the reverse shape at the second instance of the graph.

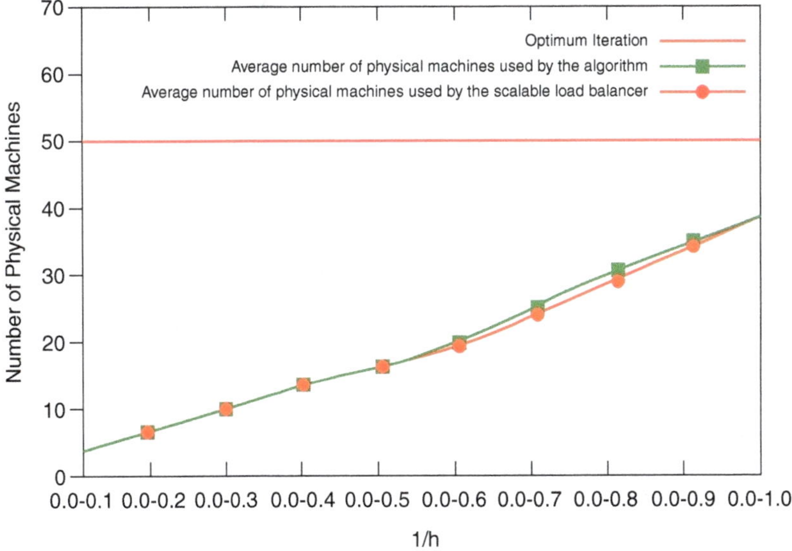

Fig. 1. Validation of optimum placement.

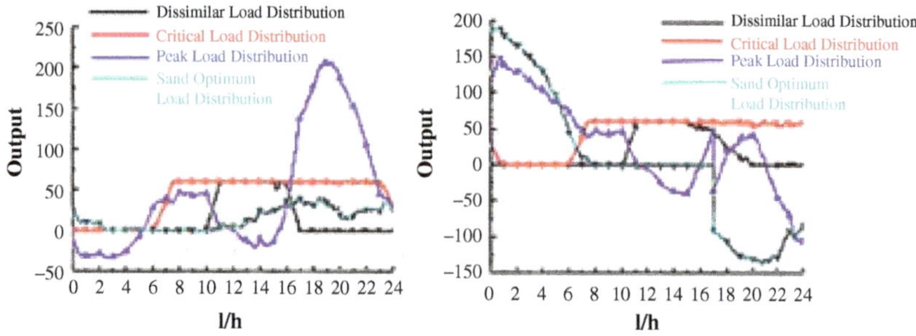

Fig. 2. Distributed changing rates of workload.

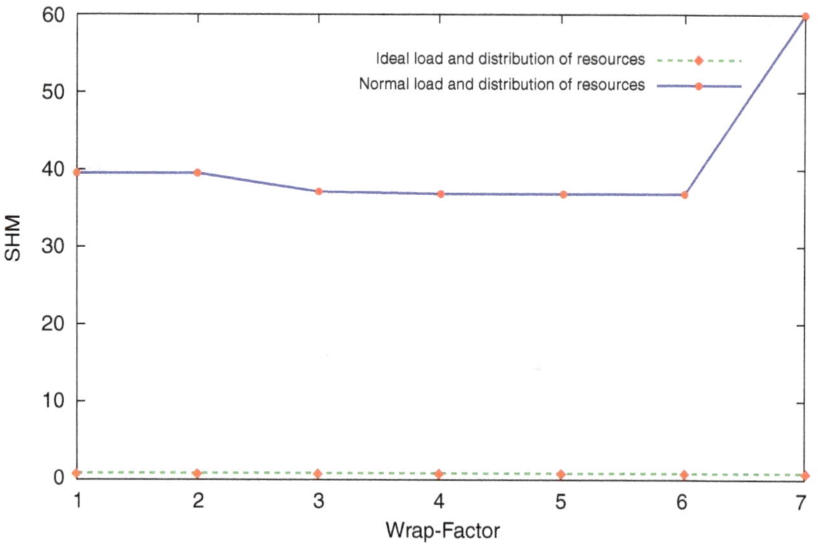

Fig. 3. Optimum mapping between wrap value and load distribution.

However, the number of initial VMs was less than that used in a load-balanced case. This indicates consolidation of the load. It is also observed that every placement plan given by any heuristic was validated against the set of constraints defined in the initial formulation of heuristics (Equations (8.4) and (8.5)) since these are the conditions that should be near optimal for any valid strategy of a virtual load-balancing scheme. We observe that all the placement plans are valid with respect to the followed dataset. In case of Figs. 3 and 4, wrap-factor vs. standard homogeneous mean (SHM) of load distribution has been considered. As an ideal and best

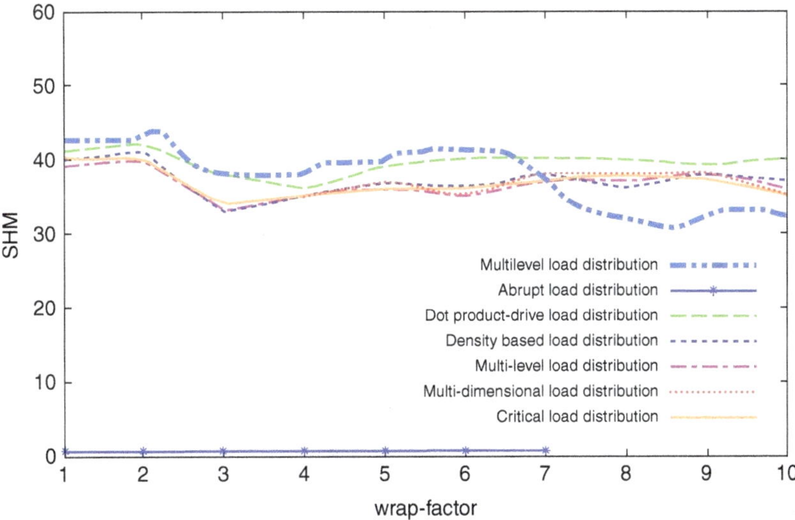

Fig. 4. Multilevel mapping between wrap value and load distribution.

Table 4. Comparison with benchmarked values.

Algorithm	No. of PMs	No. of mitigations	Time elapsed
Sandpiper	3	1	1.1
Entropy	1	3	1.5
Proposed Scalable LB	3	2	1

case, the load and distribution of resources are linear. However, in Fig. 4, multilevel mapping and abrupt load distribution are shown in bold blue marking. This behavior requires consolidation of load characteristics.

5.1 *Evaluation of performance*

The results are compared with standard algorithms and significant parametric evaluation has been conceived. The distribution of VMs under a variable load is identified with the proposed scalable load-balancing algorithm.

The proposed algorithm has been commissioned using MATLAB R2010a and Python in a open source environment on an Intel-based 2.7 Ghz PC with 8 GB RAM. The standard algorithms are compared with the proposed scalable load balancer. The time elapsed is marginally less the number of physical machines and mitigation as shown in Table 4. The plot

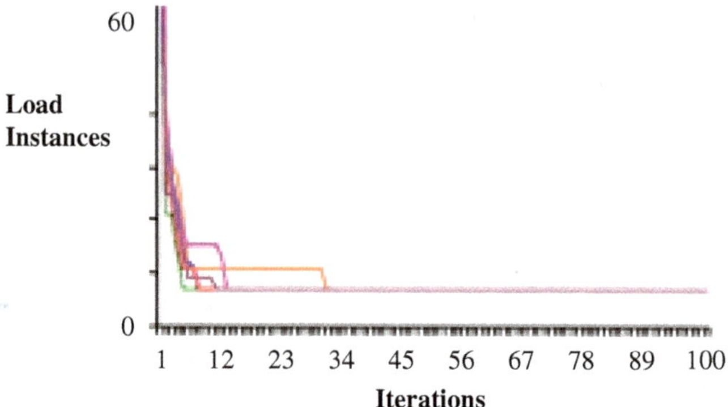

Fig. 5.   Verification scheme.

demonstrated in Fig. 5 with load instances from 0 to 60 and iterations from 0 to 100 indicates the different colors of load instances of benchmark algorithms. The plot becomes linear gradually with the value of 100.

6   Conclusion and Future Research

The objective of this chapter is to present a heuristics to investigate the optimal load-distribution strategy based on the self-organized principle. Hence, an algorithm for the optimization of cloud load balancing has been proposed, which is adaptive and supports multi-objective optimization. It is based on a hybrid approach, combining two optimization strategies using bi-level optimization and including bio-inspired algorithms in a generic manner. It therefore targets the needs of modern Big Data and IoT cloud environments, which require more sophisticated load-balancing techniques due to the high volatility of their resource consumption. The approach has been evaluated using simulations and its results are promising compared to previous works. Further research is needed to compare the approach with other previous works and to evaluate its feasibility in realistic cloud infrastructures in lab and field experiments.

References

[1]   Kleiminger, W., Mattern, F., & Santini, S. (2014). Predicting household occupancy for smart heating control: A comparative performance analysis of state-of-the-art approaches. *Energy and Buildings*, 85, 493–505.

[2]   Kleiminger, W., & Kalyvianaki, E. (2011). Peter Pietzuch balancing load in stream processing with the cloud. In *Proceedings of the 6th IEEE*

International Workshop on Self Managing Database Systems (SMDB 2011), Hannover, Germany, April.

[3] Kashefikia, M., Nematbakhsh, N., & Moghadam, R. A. (2011). Multiple ant-bee colony optimization for load balancing in packet-switched networks. *International Journal of Computer Networks & Communications*, 3(5).

[4] Javanmardi, S., Shojafar, M., Amendola, D., Cordeschi, N., Liu, H., & Abraham, A. (2014). Hybrid job scheduling algorithm for cloud computing environment. In P. Krömer *et al.* (eds.), *Proceedings of the Fifth International Conference on Innovations in Bio-Inspired Computing and Applications IBICA*, Advances in Intelligent Systems and Computing, Vol. 303, Springer.

[5] Randles, M., Lamb, D., & Taleb-Bendiab, A. (2010). A comparative study into distributed load balancing algorithms for cloud computing. In *IEEE Advanced Information Networking and Applications Workshops (WAINA)*, pp. 551–556.

[6] Florence, P., & Shanthi, V. (2014). A load balancing model using firefly algorithm in cloud computing. *Journal of Computer Science*, 10(7), 1156–1165. ISSN:1549-3636.

[7] Babu, L. D. D., & Krishna, P. V. (2013). Honey bee behavior inspired load balancing of tasks in cloud computing environments. *Applied Soft Computing Journal*, 13(5), 2292–2303.

[8] Pan, J.-S., Wang, H., & Zhao, H. (2015). Interaction artificial bee colony based load balance method in cloud computing. *Advances in Intelligent Systems and Computing*, 329.

[9] Sheeja, Y. S., & Jayalekshmi, S. (2014). Cost effective load balancing based on honey bee behaviour in cloud environment. In *IEEE 2014 First International Conference on Computational Systems and Communications (ICCSC)*, 17–18 December, Trivandrum.

[10] Mondal, B., Dasgupta, K., & Dutta, P. (2012). Load balancing in cloud computing using Stochastic hill climbing — A soft computing approach. *Procedia Technology*, 4, 783–789.

[11] Mohapatra, S., Rekha, K. S., & Mohanty, S. (2013). A comparison of four popular heuristics for load balancing of virtual machines in cloud computing. *International Journal of Computer Applications*, 68.

[12] Kim, J.-S., Nam, B., & Sussman, A. (2013). Autonomic load balancing mechanisms in the P2P desktop grid. In *Proceedings of the ACM Cloud and Autonomic Computing Conference*, 05–09 August, Miami, FL, USA.

[13] Gupta, H., & Sahu, K. (2014). Honey bee behavior based load balancing of tasks in cloud computing. *International Journal of Science and Research*, 3(6).

[14] Hefny, A. H., Khafagy, M. H., & Wahdan, A. M. (2014). Comparative study load balance algorithms for map reduce environment. *International Journal of Computer Applications*, 106(18), 41–50.

[15] Gufler, B., Augsten, N., Reiser, A., & Kemper, A. (2012). Load balancing in mapReduce based on scalable cardinality estimates. In *Proceedings of 2012*

IEEE 28th International Conference on Data Engineering, 1–5 April 2012, Washington, DC, USA.

[16] Paravastu, R., Scarlat, R., & Chandrasekaran, B. (2012). Adaptive load balancing in MapReduce using flubber. Department of Computer Science, Duke University.

[17] Atayero, A. A., Luka, M. K., & Alatishe, A. A. (2012). Neural-encoded fuzzy models for load balancing in 3GPP LTE. *International Journal of Applied Information Systems*, 4(1). ISSN:2249–0868.

[18] Bhattacharya, M. *et al.* (2016). Evolutionary optimization: A big data perspective. *Journal of Network and Computer Applications (2014)*, 59, 416–426.

[19] Buyya, R., Ranjan, R., & Calheiros, R. N. (2009). Modeling and simulation of scalable cloud computing environments and the CloudSim toolkit: Challenges and opportunities. In *Proceedings of the 7th High Performance Computing and Simulation Conference (HPCS 2009)*, 21–24 June, New York, USA, Leipzig, Germany.

[20] Radojevic, B., & Zagar, M. (2013). Analysis of issues with load balancing algorithm in hosted (cloud) environments. In *Proceedings of 34th International Convention on MIPRO*, 23–27 May, Optija, Croatia, pp. 416–420.

[21] Sundmaeker, H., Guillemin, P., Friess, P., & Woelfflé, S. European Union European Commission Directorate-General for the Information Society and Media. (2010). Vision and challenges for realizing the internet of things. Publications Office of the European Union.

[22] Brucker, P. (2007). *Scheduling Algorithms*, 5th edn., Vol. 3. Springer-Verlag, pp. 378.

[23] Amandeep, Yadav, V., & Mohammad, F. (2014). Different strategies for load balancing in cloud computing environment: A critical study. *International Journal of Scientific Research Engineering & Technology*, 3(1), 85–90. ISSN: 2278–0882.

[24] Pavlo, A., Paulson, E., Rasin, A., Abadi, D. J., DeWitt, D. J., Madden, S., Mohamed, N., Nuaimi, M. A., & Al-Jaroodi. J. (2012). A survey of load balancing in cloud computing: Challenges and algorithms. In *Proceedings of IEEE Symposium on Network Cloud Computing and Applications (NCCA)*, London pp. 137–142.

[25] Rajan, R. G., & Jeyakrishnan. V. (2013). A survey on load balancing in cloud computing environments. *International Journal of Advanced Research in Computer and Communication Engineering*, 2(12).

[26] Sonebraker. (2009). A comparison of approaches to large-scale data analysis. In *International ACM Conference on Management of Data (ACM SIGMOD)*, pp. 165–178.

[27] Florence, A., & Shanthi, V. (2014). A load balancing model using firefly algorithm in cloud computing. *Journal of Computer Science*, 10(7), 1156–1165.

[28] Cheng, S., Raja, A., & Xie, L. (2014). Dynamic multi-agent load balancing using distributed constraint optimization techniques. *International Journal Web Intelligence and Agent Systems*, 12(2), 111–138.

[29] Keskinturk, T., Yildirim, M. B., & Barut, M. (2012). An ant colony optimization algorithm for load balancing in parallel machines with sequence-dependent setup times. *Computers and Operations Research*, 39, 1225–1235.

[30] He, K., Rozner, E., Agarwal, K., Akela, A., & John, W. F. (2015). CarterPresto: Edge-based load balancing for fast datacenter networks. In *SIGCOMM'15*, 17–21, August London, United Kingdom.

[31] Sethi, S., Anupama, S., & Jena, K. (2012). Efficient load balancing in cloud computing using fuzzy logic. *IOSR Journal of Engineering*, 2(7), 65–71.

[32] Mishra, M., Das, A., Kulkarni, P., & Sahoo, A. (2012). Dynamic resource management using virtual machine migrations. *Communications Magazine*, 50(9), 34–40.

[33] Wood, T., Shenoy, P., & Arun. (2007). Black-box and gray-box strategies for virtual machine migration. In *4th USENIX Symposium on Networked Systems Design and Implementation*, 11–13 April, Cambridge, MA.

[34] Hermenier, F., Muller, G., Lorca, X., Lawall, J., & Menaud, J. M. (2009). Entropy: A consolidation manager for clusters. *5th International Conference on Virtual execution environments (VEE)*, March Washington, DC, United States.

[35] Dam, S., Mandal, G., Dasgupta, K., & Dutta, P. (2014). An ant colony based load balancing strategy in cloud computing, In M. K. Kundu *et al.* (eds.), *Advanced Computing, Networking and Informatics*, Vol. 2, Smart Innovation, Systems and Technologies 28, Springer International Publishing, Switzerland pp. 403–413.

[36] Dam, S., Mandal, G., Dasgupta, K., & Dutta, P. (2015). Genetic algorithm and gravitational emulation based hybrid load balancing strategy in cloud computing. In *Proceedings of the Third International Conference on Computer, Communication, Control and Information Technology (c3IT)*, 7–8 February, Hooghly, India.

[37] Geronimo, G. A., Uriarte, R. B., & Westphall, C. B. (2016). Towards a framework for VM organisation based on multi-objectives. *ICN 2016, The Fifteenth International Conference on Networks*, Lisbon, Portugal.

[38] Bonabeau, E., Dorigo, M., & Theraulaz, G. (1999). *Swarm Intelligence: From Natural to Artificial Systems*. Oxford University Press, New York.

[39] Mezura-Montes, E., Velzquez-Reyes, J., & Coello, C. A. (2006). A comparative study of differential evolution variants for global optimization. In *Proceedings of the 8th Annual Conference on Genetic and Evolutionary Computation*. ACM Press, New York, pp. 485–492.

[40] Kokilavani, T., & Dr. Amalarethinam, D. I. G. (2011). Load balanced min-min algorithm for static meta task scheduling in grid computing. *International Journal of Computer Applications*, 20(2).

[41] Ferreto, T. C., Netto, M. A., Calheiros, R. N., & De Rose, C. A. (2011). Server consolidation with migration control for virtualized data centers. *Future Generation Computer Systems*, 27(8), 1027–1034.

[42] Deb, K., & Sinha, A. (2010). An efficient and accurate solution methodology for bilevel multi-objective programming problems using a hybrid

evolutionary-local-search algorithm. *Evolutionary Computation Journal*, 18(3), 403–449.

[43] Elrotuba, M., & Gherbib, A. (2018). Virtual machine classification-based approach to enhanced workload balancing for cloud computing applications workload balancing for cloud computing application, In *9th International Conference on Ambient Systems, Networks and Technologies, ANT-2018, Procedia Computer Science*, 130, pp. 683–688.

[44] Kim, H. Y., & Kim, J. M. (2016). A load balancing scheme based on deep-learning in IoT. *Cluster Computing*, Springer-Verlag.

[45] Memeti, S., LiL, Pllana, S., Kolodziej, J., & Kessler, C. (2017). Benchmarking opencl, openacc, openmp and cuda: Programming productivity, performance, and energy consumption. In *Proceedings of the 2017 Workshop on Adaptive Resource Management and Scheduling for Cloud Computing, ACM*, New York, NY, USA, pp. 1–6.

[46] Grzonka, D., Jakbik, A., Koodziej, J., & Pllana, S. (2017). Using a multi-agent system and artificial intelligence for monitoring and improving the cloud performance and security. *Future Generation Computer Systems*, https://doi.org/10.1016/j.future.2017.05.046.

[47] Lyu, H., Li, P., Yan, R., & Luo, Y. (2016). Load forecast of resource scheduler in cloud architecture. *International Conference on Progress in Informatics and Computing (PIC)*, 23–25 December, Shanghai, China.

[48] Roshdy, A., Gaber, A., Hantera, F., & ElSebai, M. (2018). Mobility load balancing using machine learning with case study in live network. In *IEEE 2018 International Conference on Innovative Trends in Computer Engineering (ITCE)*, 19–21 February.

[49] Qi, J., Hongsheng, X., Baoqun, Y., & An, X. (2008). Event-driven dynamic load balancing strategy for streaming media clustered server systems. In *Chenfeng Proceedings of the 27th Chinese Control Conference*, 16–18 July, Kunming, Yunnan, China.

[50] Yagoubi, B., & Meddeber, M. (2010). Distributed load balancing model for grid computing. *Journal of Computer Science*, 12, 43–60.

[51] Indhumathi, V., & Nasira, G.M. (2016). Service oriented architecture for load balancing with fault tolerant in grid computing. In *IEEE International Conference on Advances in Computer Applications (ICACA)*, pp. 313–317.

[52] Naqvi, S. A. A., Javaid, N., Butt, H., Kamal, M. B., Hamza, A., & Kashif, M. (2018). Metaheuristic optimization technique for load balancing in cloud-fog environment integrated with smart grid. In *International Conference on Network-Based Information Systems*, Springer, Cham, pp. 700–711.

[53] Katyal, M., & Mishra, A. (2014). A comparative study of load balancing algorithms in cloud computing environment. arXiv:1403.6918.

[54] Singh, A., Juneja, D., & Malhotra, M. (2015). Autonomous agent based load balancing algorithm in cloud computing. *Procedia Computer Science*, 45, 832–841.

[55] Milani, A. S., & Navimipour, N. J. (2016). Load balancing mechanisms and techniques in the cloud environments: Systematic literature review and future trends. *Journal of Network and Computer Applications*, 71, 86–98.

[56] Ghomi, E. J., Rahmani, A. M., & Qader, N. N. (2017). Load-balancing algorithms in cloud computing: A survey. *Journal of Network and Computer Applications*, 88, 50–71.

[57] Liu, Q., Cai, W., Shen, J., Fu, Z., Liu, X., & Linge, N. (2016). A speculative approach to spatial-temporal efficiency with multi-objective optimization in a heterogeneous cloud environment. *Security and Communication Networks*, 9(17), 4002–4012.

[58] Singh, S., & Chana, I. (2016). A survey on resource scheduling in cloud computing: Issues and challenges. *Journal of Grid Computing*, 14(2), 217–264.

[59] Rana, M., Bilgaiyan, S., & Kar, U. (2014, July). A study on load balancing in cloud computing environment using evolutionary and swarm based algorithms. In *Control, Instrumentation, Communication and Computational Technologies (ICCICCT)*, pp. 245–250.

[60] Ragmani, A., El Omri, A., Abghour, N., Moussaid, K., & Rida, M. (2016). A performed load balancing algorithm for public Cloud computing using ant colony optimization. In *2nd International Conference on Cloud Computing Technologies and Applications (CloudTech)*, pp. 221–228.

[61] Gabi, D., Ismail, A. S., & Zainal, A. (2015). Systematic review on existing load balancing techniques in cloud computing. *International Journal of Computer Applications*, 125(9).

[62] Ashwin, T. S., Domanal, S. G., & Guddeti, R. M. R. (2014). A novel bio-inspired load balancing of virtual machines in cloud environment. In *IEEE International Conference on Cloud Computing In Emerging Markets (Ccem)*, pp. 1–4.

[63] Farahnakian, F., Ashraf, A., Pahikkala, T., Liljeberg, P., Plosila, J., Porres, I., & Tenhunen, H. (2015). Using ant colony system to consolidate VMs for green cloud computing. *IEEE Transactions on Services Computing*, 8(2), 187–198.

[64] Ghumman, N. S., & Kaur, R. (2015). Dynamic combination of improved max-min and ant colony algorithm for load balancing in cloud system. In *Computing, Communication and Networking Technologies (ICCCNT)*, pp. 1–5.

[65] Schedel, A., & Brune, P. (2017). Predicting response time-related quality-of-service outages of PaaS cloud applications by machine learning. In *International Conference on Mobile, Secure, and Programmable Networking*, Springer, Cham, pp. 155–165.

[66] Gomez, C. A., Shami, A., & Wang, X. (2018). Machine learning aided scheme for load balancing in dense IoT networks. *Sensors*, 18, 3779.

[67] Walsh, A. E. (2002). *Uddi, Soap, and Wsdl: The Web Services Specification Reference Book*. Prentice Hall. Professional Technical Reference A©2002 ISBN:0130857262.

[68] Cui, X., Huang, X., Ma, Y., & Meng, Q. (2019). A load balancing routing mechanism based on SDWSN in smart city. *Electronics*, 8(3).

[69] Vadacchino, V. (2016). From smart grid to smart city EXPO Milano 2015 best practice. Presentation.

[70] Sliwa, B., Falkenberg, R., & Wietfeld, C. (2017). A simple scheme for distributed passive load balancing in mobile ad-hoc networks. In *IEEE 85th Vehicular Technology Conference (VTC Spring)*, doi: 10.1109/VTC-Spring.2017.8108553.

Chapter 9

A User-Demand Privacy-Preserving Framework Based on Association Rules and Differential Privacy in Social Networks

Chunliu Yan*, Ziyi Ni*, Bin Cao*,‡, Rongxing Lu†,
Shaohua Wu* and Qinyu Zhang*
*Harbin Institute of Technology (Shenzhen),
Shenzhen, Guangdong, P. R. China, 518055
†University of New Brunswick,
Fredericton, NB, Canada, E3B 5A3
‡caobin@hit.edu.cn

1 Introduction

The past several years have demonstrated that people deeply rely on online services, such as shopping, transportation and medical services. The increasing applications of location-based services (LBSs) indeed make our lives more convenient. However, LBSs also lead to significant privacy leakage. On the one hand, location information of a single point is sensitive. On the other hand, location trajectories can excavate more sensitive information despite location. In this regard, location privacy-preserving mechanisms have been extensively studied, such as dummy, spatial confusion [1, 2], encryption technology [3, 4], differential privacy (DP) [5, 6] and references therein.

Inspired by the famous marketing strategy of "beer and diaper" [7], we incorporate the association rule and differential privacy into our system model and mechanism design. Consider a case in which a single mom needs to search for stores selling diapers for her illegitimate child. Considering that she is unwilling to expose that she has a child, i.e., keeping no search history revealing her child, how can she find the stores using her smartphone, i.e., how to protect this privacy led by user demand disclosure? Recall that beer and diaper can be linked in the essence of popular shopping habits. According to this interesting association relationship, it is more likely that the store selling beer also places diapers around. Therefore, it is reasonable to search for beer rather than diaper, and beer itself is less sensitive than diapers. Nevertheless, the search results may not satisfy her demand, i.e., only beer, which we deem as the quality of service (QoS) loss. With the aim to address user demand privacy in more generalized application scenarios, this work focuses on designing a framework and techniques which can strike a balance between user demand privacy and QoS loss. To this end, we propose UMBRELLA: user demand privacy preserving based on association rules and differential privacy in social networks.

2   Related Work

Although there is little related research on the above-mentioned problem, methodologies adopted by location privacy can provide useful references. In this section, we review some related works regarding location privacy.

2.1   *Spatial cloaking*

Spatial cloaking is a popular scheme, which transfers the real location to cloaked regions based on K-anonymity. Generally speaking, noise is added to the real location so that the attacker gets a fuzzy position information. The types of noise play an important role in both privacy and QoS, e.g., stronger noise leads to better privacy and less QoS. Gruteser [8] first applied the K-anonymity to location privacy. Each user is indistinguishable from at least $K-1$ users at a certain time and space. Hence, the attacker cannot identify the target users and infer their real locations. However, this scheme may be invalid in some cases due to limitations of the K-anonymity. To tackle this issue, Mokbel proposed a Casper model [9]. In this scheme, based on classical K-anonymity, a privacy parameter A_{min} was defined to restrict the cloaking region. A trusted third party was introduced to link the users and the servers so as to form a fuzzy space. To avoid potential risks in the third party, Shao further proposed a third-party-free protocol, which can provide full privacy [10]. Thanks to the dependence of locations over time, Montazeri defined the perfect location privacy and proposed a mathematical framework for location privacy based on information theory [11]. This framework models the real movements of users over time by using Markov chains and confuses the adversary by using anonymization techniques. Perfect location privacy is achievable under some specific conditions.

2.2 *Dummy and query privacy*

Dummy is also a typical method in location privacy. Different from some classical schemes [12, 13], Niu proposed DLS to achieve K-anonymity [14]. Considering that the adversary has side information, each single location to be queried is usually with a querying probability due to the searching history. In this regard, entropy is used to measure the degree of anonymity, turning out that achieving the maximum entropy brings the best privacy. Although users may avoid distributing their information in some sensitive locations, an adversary could infer their privacy from spatial–temporal correlations and historical information. In order to solve this problem, a novel metric to quantify the context-aware location privacy has been proposed by Zhang *et al.* [15]. They designed two algorithms to calculate the correlation among different locations. As for the scenario that an adversary can predict the demographic characteristic of users from the shared locations, based on context information, Li proposed SmartMask [16] to learn privacy preferences in different contexts. In LBSs, query privacy is also an important and sensitive issue. Spatial cloaking lacks the capability of privacy-preserving in query process, and the service is based on the cloaking regions rather than the real location. Song proposed a system called anonymity of motion vectors to provide anonymity for spatial query [17]. This system can minimize cloaking regions by predicting users' movement based on motion vectors. In continuous LBSs, Pingle proposed a third-party-free scheme based on query perturbation [18], which still works even when the users' identity was revealed.

2.3 *PIR and encryption technology*

In addition to anonymizers in LBS, a framework to support private location-dependent queries based on private information retrieval has been proposed by Ghinita *et al.* [19]. Lu *et al.* further proposed a fine-grained privacy-preserving LBS scheme for mobile devices. The framework employs a ciphertext policy anonymous attribute-based encryption technique to achieve location privacy [20]. Moreover, Lu *et al.* proposed a privacy-preserving data dissemination protocol by utilizing the paillier cryptosystem and its homomorphic properties [21]. In addition, a framework APPLET to protect locations and location-aware recommendation results was proposed by Ma *et al.* [3]. This framework allows us to securely compute the similarities of venue through paillier encryption and to predict the recommendation results based on Paillier, commutative and comparable encryption.

2.4 *Differential privacy*

DP was proposed by Dwork in 2006 [22]. It essentially translates the query results of the dataset into the same distribution, so that two adjacent datasets lead to the same result. Due to the superiorities of DP, it has been widely studied and applied in data publishing and data mining. Note that DP is more suitable to protect multi-user information aggregation, while it is not appropriate to address privacy issue for an individual user.

In practical scenarios, both the user and the adversary aim to optimize the strategies and obtain their benefits. A pioneering work by Andres *et al.* presented a geo-indistinguishability notion based on generalized DP [5]. The geo-indistinguishability can be achieved by adding two-dimensional Laplace noise to the real location. Different from [6, 23], the focus of Shokris work is on applying linear programming to balance the privacy and quality. In addition, Kellaris *et al.* applied DP to improve the effectiveness, so that both privacy and quality are guaranteed [24].

3 System Model

This section provides a description of the system model and the design goal.

3.1 *Description*

Our model can be mainly divided into three layers, namely, entity layer, strategy layer and evaluation layer, as depicted in Fig. 1.

- Entity layer mainly involves users, service providers and attackers. A user sends a service request to the service provider, and the service provider returns the result to the user. Attackers directly attack the service providers to obtain the query content. It is possible that the service provider is untrustworthy. In addition, the attacker may also intercept the communication between users and service providers.
- Strategy layer mainly establishes a mathematical model by combining association rules and DP technology. The quantitative indicators of

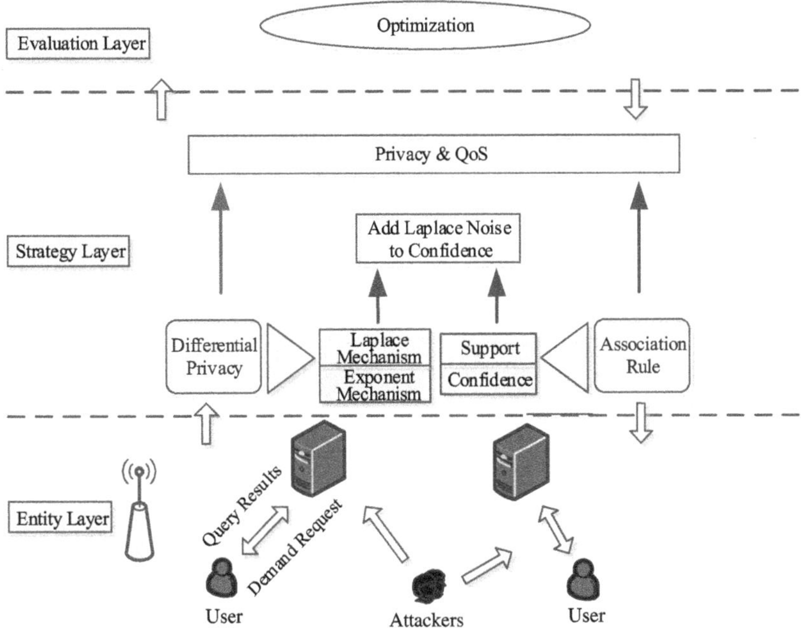

Fig. 1. System framework.

association rules are the support degree and confidence level. The DP is performed via the Laplace mechanism, i.e., adding noise to the confidence level.

- Evaluation layer analyzes and optimizes system performance in terms of mathematical formulation and optimization, e.g., game theory and convex optimization, to strike a balance between privacy and QoS.

The main focus of our work is on the strategy layer and the evaluation layer.

3.2 *Design goal*

Given the system modeled above, our design goal is to develop an effective demand privacy-preserving scheme in a social network. Specially, the following two objectives should be achieved:

- The demand privacy should be guaranteed.
- The balance between privacy and QoS should be ensured, given that the system supports various privacy preferences.

3.3 *Scenario in smart city*

At Wal-Mart stores in the United States, beer and diapers are usually displayed in the same shopping rack. This move is also convenient for people to shop while increasing the sales of the two. After investigation, young women in the United States often ask their husbands to buy diapers for their baby after work. These dads will choose beer after they choose the diapers. After supermarket staff discovered this phenomenon, they put the diapers and the beer in the same product column, which not only increased the sales profit of the two but also facilitated people's shopping. On the other hand, if a single mother wants to find out which supermarket sells diapers, but does not want others to know she has a baby, she can try to search which supermarket sells beer, which not only ensures that the single mother's privacy is not leaked but also provides the corresponding demand services.

4 Preliminaries

In this section, we outline the association rules and the DP, which will serve as the basis of UMBRELLA.

4.1 *Association rules*

Suppose $I = \{i_1, i_2, i_3, \ldots, i_n\}$ is a set of n different items. The sets consisting of some elements of I are its subsets [25]. A transaction is denoted by $t = < ID, X >$, where ID is the unique identifier and X is a set representing a transaction. The form of the association rule is given as follows: $X \rightarrow Y$, where both X and Y are item sets, $X \cap Y = \phi$, and ϕ is an empty set, showing X can infer to Y. X is named the condition of the rule, and Y is called the conclusion. Let S be the support of the rule $X \rightarrow Y$ representing the ratio of X and Y to all transactions in the database, i.e.,

$$S(X,Y) = P_r(XY) = \frac{\text{number}(XY)}{\text{number}(\text{AllSamples})}. \tag{9.1}$$

If the rule is $X, Y \rightarrow Z$, S is obtained as

$$S(X,Y,Z) = P_r(XYZ) = \frac{\text{number}(XYZ)}{\text{number}(\text{AllSamples})}. \tag{9.2}$$

Define C as the confidence of the rule $X \rightarrow Y$ which represents the ratio of X and Y to transactions Y, i.e.,

$$C(X \rightarrow Y) = P_r(X|Y) = \frac{P_r(XY)}{P_r(Y)}. \tag{9.3}$$

If the rule is $X, Y \rightarrow Z$, C is calculated as

$$C(XY \rightarrow Z) = P_r(Z|YX) = \frac{P_r(XYZ)}{P_r(XY)}. \tag{9.4}$$

Consider that users can specify their minimum support threshold (MST) and minimum confidence threshold (MCT). $X \rightarrow Y$ is a strong association rule if and only if $S(X) \geq$ MST and $C(X \rightarrow Y) \geq$ MCT.

In our system model, we assume that strong association rules are available from advanced data mining techniques [26] and focus on the confidence issues. Note that under strong association rules, different confidence levels play a considerable impact on the relevance of demand.

4.2 *Differential privacy*

The rationale behind using DP is to further protect privacy. In our work, we define the privacy leakage boundary to restrict the leakage level. The principle of DP is to ensure that there are different demands corresponding to the same replacement. In this context, the attacker cannot distinguish the real demand from fuzzy results. In a nutshell, for any two datasets D and D' differing only in a single item, with the same output O under a privacy protection function g, we have

$$\frac{Pr\left\{g(D) \in O\right\}}{Pr\left\{g(D') \in O\right\}} \leq \exp(\epsilon), \tag{9.5}$$

wherein the privacy budget parameter ϵ controls the trade-off between the privacy and utility.

In general, there are two mechanisms for the realization of DP, namely, the Laplace mechanism and the exponential mechanism. In our model, the Laplace mechanism is used to add noise to the confidence. The probability distribution function of Laplace is given by

$$f(x|\mu, b) = \frac{1}{2b} e^{-\frac{|x-\mu|}{b}}, \tag{9.6}$$

where μ and $b = \frac{1}{\epsilon}$ represent the position parameter and the scale parameter, respectively.

5 Proposed UMBRELLA Scheme

Inspired by [23], we propose a demand privacy-preserving scheme based on association rules and the DP in social network (UMBRELLA).

A real demand is denoted by s, and S is a set of all possible items consisting of s. Let prior knowledge ω be the probability distribution of s, which is written as

$$\omega(s) = P_r\{S = s\}. \tag{9.7}$$

The prior knowledge ω is obtained from historical user demands. When the user demands change, the probability distribution changes accordingly. Thus, ω depending on time should be dynamically adapted.

5.1 *Protection mechanism*

We assume that the real user demand is s, and the replacement is $o \in O$. Suppose the replacement is available to attackers. A probabilistic protection mechanism is to replace real demand with other replacements, i.e.,

$$p(o|s) = P_r\{O = o|S = s\}. \tag{9.8}$$

Consider $O = S$, and the above probabilistic mechanism can be deemed as a channel between a user and an attacker. The greater the noise of the channel is, the higher the privacy is, while QoS decreases accordingly.

5.2 *QoS issue*

Since the real demand is replaced by other items, the strong association rules will also lead to QoS loss. Let $c(o, s)$ be the confidence between s and o,[1] and denote $\log_2(1/c(o, s))$ as the confidence distance to reflect QoS loss. We add the Laplace noise to the $c(o, s)$, and the confidence with noise is given by $c_L(o, s)$. Basically, the greater the confidence, the lesser the QoS loss, and vice versa. QoS loss depends on prior knowledge, protection mechanism and the confidence distance. There is no QoS loss as a result of

---

[1]Both the uppercase C and the lowercase c represent confidence.

which $c(o, s)$ equals one. In this regard, similar to [23], QoS loss is written as

$$\sum_s \omega(s) \sum_o p(o|s) \cdot \log_2(1/c_L(o, s)).$$ (9.9)

Particularly, $c(o, s)$ equaling to 1 shows that the real demand is replaced by itself, hence there is no QoS loss. We assume that users accept a maximum QoS loss, $\text{loss}_{\max}$, caused by sharing replacements instead of real demand.

5.3 *Attackers inference mechanism*

An attacker aims to obtain a possible demand $\hat{s}$ through o by the following inference mechanism:

$$q(\hat{s}|o) = P_r \{S = \hat{s}|O = o\}.$$ (9.10)

5.4 *Distortion privacy*

We define the inference distance function as $\log_2(1/c(\hat{s}, s))$. Given an inference function q, user's distortion privacy is given as follows [6]:

$$\sum_o p(o|s) \sum_{\hat{s}} q(\hat{s}|o) \cdot \log_2(1/c(\hat{s}, s)),$$ (9.11)

reflecting the attacker's inference error.

Statistically, the expected distortion privacy is

$$\sum_s \omega(s) \sum_o p(o|s) \sum_{\hat{s}} q(\hat{s}|o) \cdot \log_2(1/c(\hat{s}, s)).$$ (9.12)

6 Solution

In this section, we investigate the solution issues of the proposed UMBRELLA scheme, which consists of the following three parts: game theory analysis, optimal attack vs. Bayesian attack and privacy preference.

6.1 *Game theory analysis*

To strike a balance between privacy and QoS, game theory is utilized to achieve the optimal strategies. Note that the attacker adapts his/her strategy according to that of the user, and it is reasonable to formulate the problem as a two-tied game, namely, Stackelberg game. In this regard, the user is the leader, and the attacker is a follower, respectively, in the game. To tackle the optimization simply yet efficiently, we leverage zero-sum Stackelberg to analyze this problem.

As the user and the attacker conflict with each other in terms of privacy, it can be further formulated as a zero-sum game. We assume that users accept a maximum QoS loss, $\text{loss}_{\max}$, caused by sharing replacements instead of real demand. Let P and Q be strategy spaces for the user and the attacker, respectively.

$$P = \{p(o_1|s), p(o_2|s), \ldots, p(o_m|s)\},$$

$$\sum_i p(o_i|s) = 1, \quad \forall s \in S : p(o_i|s) \geq 0, \quad \forall o_i \in O, \tag{9.13}$$

and

$$Q = \{q(\hat{s}_1|o), q(\hat{s}_2|o), \ldots, q(\hat{s}_n|o)\},$$

$$\sum_j q(\hat{s}_j|o) = 1, \quad \forall o \in O : q(\hat{s}_j|o) \geq 0, \quad \forall \hat{s}_j \in S. \tag{9.14}$$

In this Stackelberg game, the adversary knows prior knowledge $\omega(s)$ and probability distribution $p(o|s)$, and only actual demand s is not known.

We compute user's optimal strategy of protection mechanism $p(\cdot)$ and adversary's optimal strategy of inference attack $q(\cdot)$, given $\omega(s)$, $\log_2(1/c(\hat{s}, s))$, $\log_2(1/c_L(o, s))$ and $\text{loss}_{\max}$. Specifically, we establish two linear programs for the user and the attacker. Let π be the user demand privacy, i.e.,

$$\pi = \sum_{s,o,\hat{s}} \omega(s) \cdot p(o|s) \cdot q(\hat{s}|o) \cdot \log_2(1/c(\hat{s}, s)). \tag{9.15}$$

- Optimal strategy for the user,

$$\max_p \pi, \tag{9.16}$$

$$\text{s.t.} \sum_{s,o} \omega(s) \cdot p(o|s) \cdot \log_2(1/c_L(o, s)) \leq \text{loss}_{\max}, \tag{9.17}$$

$$\sum_o p(o|s) = 1, \quad \forall s, \tag{9.18}$$

$$p(o|s) \geq 0, \quad \forall s, o. \tag{9.19}$$

- Optimal strategy for the attacker,

$$\min_{q} \pi, \tag{9.20}$$

$$\text{s.t.} \sum_{\hat{s}} q(\hat{s}|o) = 1, \quad \forall o, \tag{9.21}$$

$$q(\hat{s}|o) \geq 0, \quad \forall \hat{s}, o. \tag{9.22}$$

Nash equilibrium exists and is unique, and the detailed steps are as follows.

Step 1: More precisely, the optimal protection strategy p^* corresponds to the optimal attack strategy q^*. The optimal attack strategy q^* minimizes the user demand privacy, and the demand privacy is

$$\sum_{s,o,\hat{s}} \omega(s) \cdot p(o|s) \cdot q^*(\hat{s}|o) \cdot \log_2(1/c(\hat{s}, s))$$

$$= \sum_{o} \min_{q(\cdot|o)} \sum_{\hat{s}} q(\hat{s}|o) \sum_{s} \omega(s) \cdot p(o|s) \cdot \log_2(1/c(\hat{s}, s)). \tag{9.23}$$

Obviously, $\min_{q(\cdot|o)} \sum_{\hat{s}} q(\hat{s}|o) \sum_{s} \omega(s) \cdot p(o|s) \cdot \log_2(1/c(\hat{s}, s))$ is an average of $\sum_{s} \omega(s) \cdot p(o|s) \cdot \log_2(1/c(\hat{s}, s))$; hence, it must be greater than or equal to the minimum of it for a particular $\hat{s}$,

$$\min_{q(\cdot|o)} \sum_{\hat{s}} q(\hat{s}|o) \sum_{s} \omega(s) \cdot p(o|s) \cdot \log_2(1/c(\hat{s}, s))$$

$$\geq \min_{\hat{s}} \sum_{s} \omega(s) \cdot p(o|s) \cdot \log_2(1/c(\hat{s}, s)). \tag{9.24}$$

Moreover, the attacker minimizes the privacy by using q^* in all mixed strategies.

Step 2: Define $q'(\hat{s}|o)$ as a probability distribution function that represents one particular inference attack. Specially, $q'(\hat{s}|o)$ equals one if

$$\hat{s} = \operatorname{argmin} \sum_{s} \omega(s) \cdot p(o|s) \cdot \log_2(1/c(\hat{s}, s)). \tag{9.25}$$

We know that $q'(\hat{s}|o)$ as a pure strategy, and the minimum value that is optimized for all mixed strategies is less than or equal to that for pure strategy. We have

$$\min_{q(.|o)} \sum_{\hat{s}} q(\hat{s}|o) \sum_{s} \omega(s) \cdot p(o|s) \cdot \log_2(1/c(\hat{s}, s))$$

$$\leq \sum_{\hat{s}} q'(\hat{s}|o) \sum_{s} \omega(s) \cdot p(o|s) \cdot \log_2(1/c(\hat{s}, s)). \tag{9.26}$$

and

$$\sum_{\hat{s}} q'(\hat{s}|o) \sum_{s} \omega(s) \cdot p(o|s) \cdot \log_2(1/c(\hat{s}, s))$$

$$= \min_{\hat{s}} \sum_{s} \omega(s) \cdot p(o|s) \cdot \log_2(1/c(\hat{s}, s)). \tag{9.27}$$

Therefore,

$$\min_{q(.|o)} \sum_{\hat{s}} q(\hat{s}|o) \sum_{s} \omega(s) \cdot p(o|s) \cdot \log_2(1/c(\hat{s}, s))$$

$$\leq \min_{\hat{s}} \sum_{s} \omega(s) \cdot p(o|s) \cdot \log_2(1/c(\hat{s}, s)). \tag{9.28}$$

Step 3: Combining (9.24) and (9.28), we have

$$\min_{q(.|o)} \sum_{\hat{s}} q(\hat{s}|o) \sum_{s} \omega(s) \cdot p(o|s) \cdot \log_2(1/c(\hat{s}, s))$$

$$= \min_{\hat{s}} \sum_{s} \omega(s) \cdot p(o|s) \cdot \log_2(1/c(\hat{s}, s)). \tag{9.29}$$

Furthermore, combining (9.23) and (9.29), we have

$$\sum_{s,o,\hat{s}} \omega(s) \cdot p(o|s) \cdot q^*(\hat{s}|o) \cdot \log_2(1/c(\hat{s}, s))$$

$$= \sum_{o} \min_{\hat{s}} \sum_{s} \omega(s) \cdot p(o|s) \cdot \log_2(1/c(\hat{s}, s)). \tag{9.30}$$

Naturally, we convert the linear programming of user's optimal strategy as

$$\max \sum_{o} \min_{\hat{s}} \sum_{s} \omega(s) \cdot p(o|s) \cdot \log_2(1/c(\hat{s}, s)), \tag{9.31}$$

$$\text{s.t.} \sum_{s,o} \omega(s) \cdot p(o|s) \cdot \log_2(1/c_L(o,s)) \leq \text{loss}_{\max}, \tag{9.32}$$

$$\sum_o p(o|s) = 1, \quad \forall s, \tag{9.33}$$

$$p(o|s) \geq 0, \quad \forall s, o. \tag{9.34}$$

Mathematically,

$$\pi(p_i^*, p_{-i}^*, q^*) \geq \pi(p_i, p_{-i}^*, q^*), \tag{9.35}$$

where $p_i^* = p(o_i|s)$ and $p_{-i}^* = P - p_i^*$.

$$\pi(q_j^*, q_{-j}^*, p^*) \leq \pi(q_j, q_{-j}^*, p^*), \tag{9.36}$$

where $q_j^* = q(\hat{s}_j|o)$ and $q_{-j}^* = Q - q_j^*$.

To summarize, the strategy profile (p^*, q^*) is a strict Nash equilibrium.

6.2 *Privacy preference*

Consider a user with N privacy preferences, i.e., demand privacy can be divided into $0, 1, \ldots, N-1$ levels. To satisfy privacy preferences, the system needs to be adaptive. In our model, it is necessary to change the element numbers in O to guarantee the user demand privacy preference under a certain QoS loss.

7 Performance Evaluation and Analysis

In this section, we evaluate and analyze the performance of the proposed scheme in terms of privacy, QoS and adaptiveness of the system.

7.1 *Scenario 1*

In this case, we adopt the zero-sum Stackelberg game method to perform an effective and robust solution. Ten different s of a user and 15 o are selected. Suppose all $\omega(s)$ are available to attackers. Both $c(o, s)$ and $c(\hat{s}, s)$ are set between 0.1 and 1. The privacy budget parameter ϵ equals 0.2 and the Laplace noise parameter μ equals 0. The minimum privacy d_m ranges from 0 to 1.

In Fig. 2, demand privacy increases with QoS loss at an initial stage. Further, the privacy remains unchanged because it is a equilibrium state for both the user and the attacker. The performance of UMBRELLA with and without DP is shown in Figs. 4 and 5. We further account

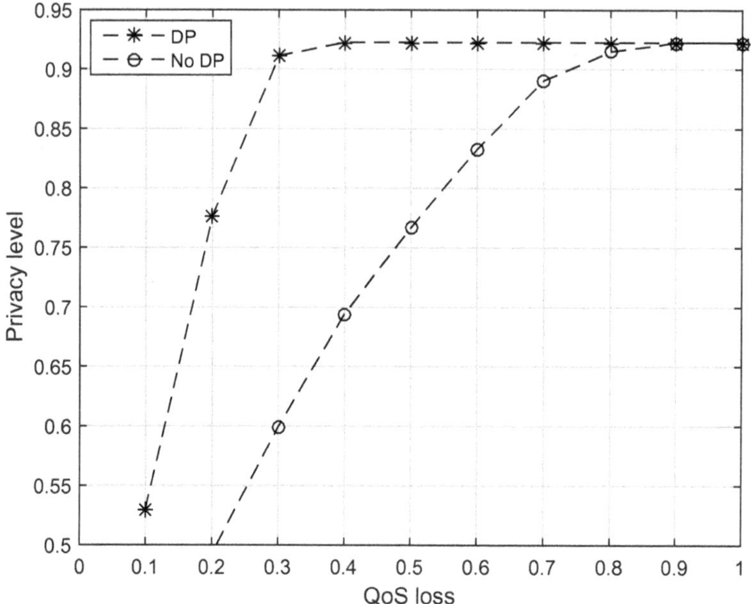

Fig. 2. Privacy vs. QoS, with and without DP.

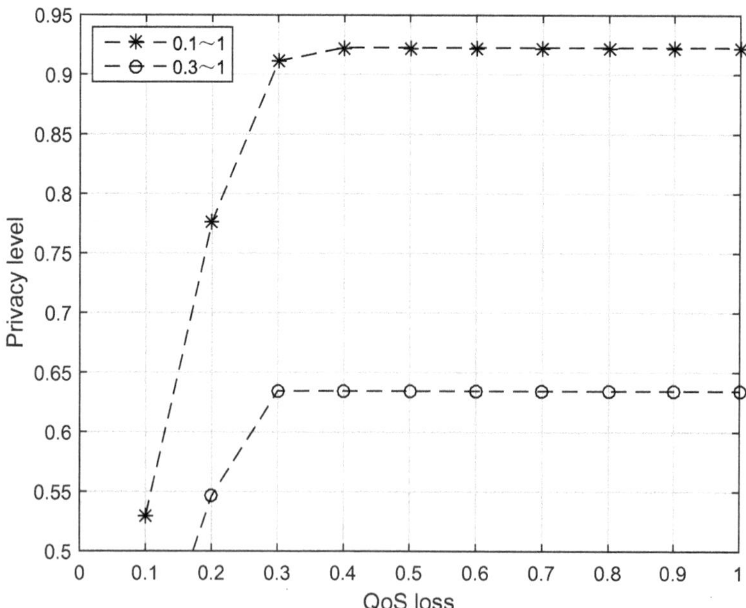

Fig. 3. Privacy vs. QoS loss under different confidence levels.

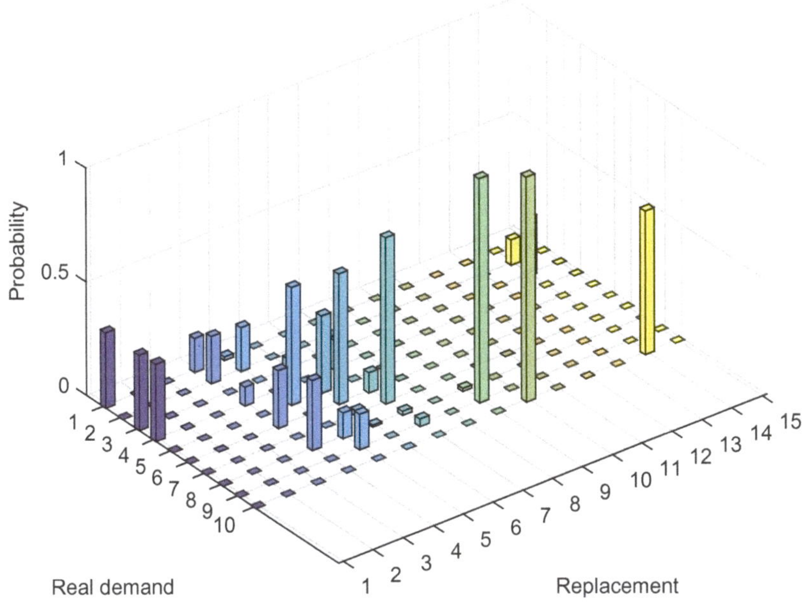

Fig. 4. Probability distribution of replacements without DP.

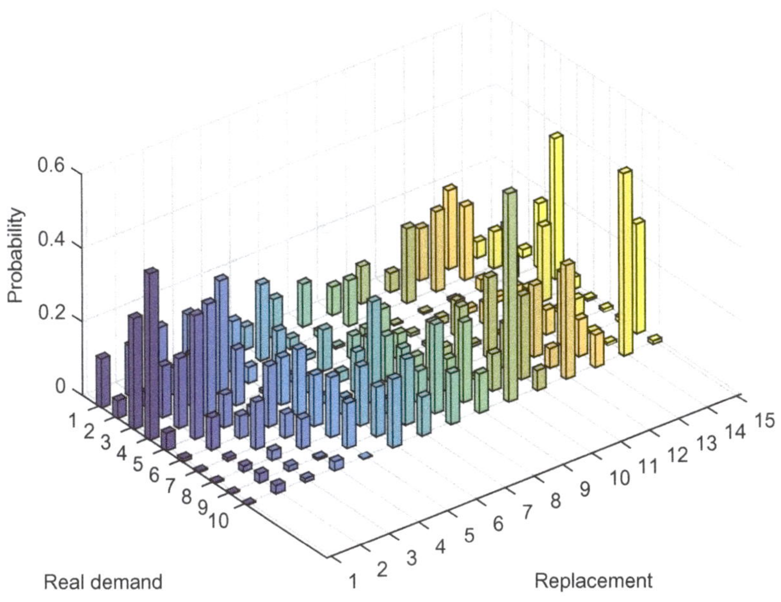

Fig. 5. Probability distribution of replacements with DP.

for the performance of UMBRELLA with DP. The variances in Figs. 4 and 5 are $\sigma_1^2 = 0.1646$ and $\sigma_2^2 = 0.1298$, respectively. Obviously, the probability distribution is more concentrated with DP. This result leads to the conclusion that the probabilities of an o mapping into multiple s are very close, hence inference probability is smaller with DP, and the privacy with DP outperforms that without DP.

Inference distance function $\log_2(1/c(\hat{s}, s))$ also affects user demand privacy, notably, demand privacy is better protected as $c(\hat{s}, s)$ decreases in Fig. 3. Lesser confidence leads to a greater inference distance function, apparently, and the privacy becomes larger according to the formula (9.11).

7.2 *Scenario 2*

As users have a certain privacy preference, the system needs to change the number of o to meet different privacy requires. In our experiment, demand privacy is proportional to the number of o under a same QoS loss. Obviously, different privacy levels can be guaranteed in Fig. 6.

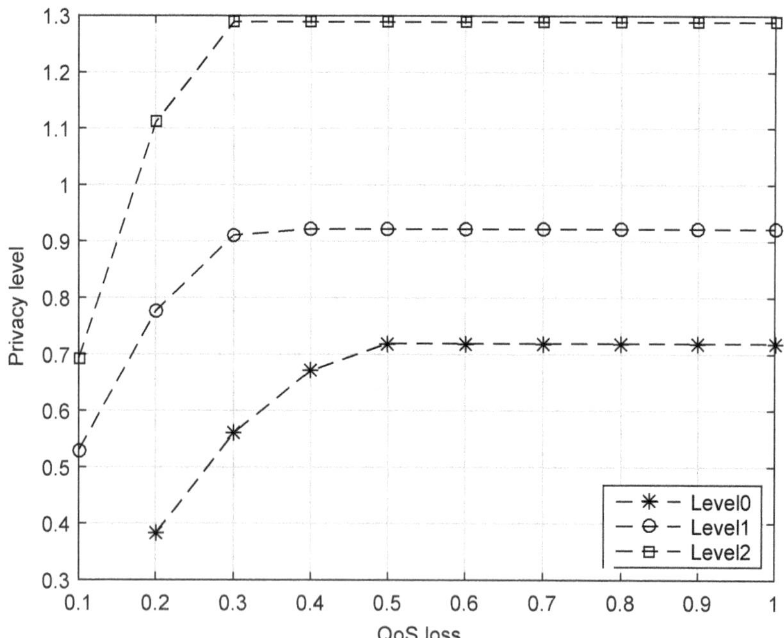

Fig. 6. Privacy vs. QoS loss under different privacy preference levels.

8 Conclusions

In this chapter, we have proposed a novel demand privacy-preserving scheme — UMBRELLA. Based on the association rules and DP, UMBRELLA is identified to be not only capable of maximizing QoS but also capable of striking a balance between the user and the attacker in terms of the demand privacy. The system can also select the number of replacements dynamically to protect privacy according to privacy preferences. Furthermore, we provide comparative experiments to demonstrate the security strength and the privacy-preserving ability with and without DP. In the future, we will study the protection of user demand trajectory and further protect demand privacy.

References

[1] Peng, T., Liu, Q., & Wang, G. J. (2014). Enhanced location privacy preserving scheme in location-based services. *IEEE Systems Journal*, 11(99), 1–12.

[2] Shahid, A. R., Jeukeng, L., Zeng, W., Pissinou, N., Iyengar, S., Sahni, S., & Varela-Conover, M. (2017). PPVC: Privacy preserving voronoi cell for location-based services. In *Proceedings of International Conference on Computing, Networking and Communications*, Silicon Valley, pp. 351–355.

[3] Ma, X. D., Li, H., Ma, J. F., Jiang, Q., Gao, S., Xi, N., & Lu, D. (2017). APPLET: A privacy-preserving framework for location-aware recommender system. *Science*, 60, 092101.

[4] Zhu, H., Lu, R. X., Huang, C., Chen, L., & Li, H. (2016). An efficient privacy-preserving location-based services query scheme in outsourced cloud. *IEEE Transactions on Vehicular Technology*, 65, 7729–7739.

[5] Andres, M. E., Bordenabe, N. E., Chatzikokolakis, K., & Palamidessi, C. (2013). Geo-indistinguishability: Differential privacy for location-based systems. In *Proceedings of the 2013 ACM SIGSAC Conference on Computer and Communications Security*, Berlin, pp. 901C914.

[6] Shokri, R. (2015). Privacy games: Optimal user-centric data obfuscation. In *Proceedings of the 15th Privacy Enhancing Technologies*, Philadelphia, pp. 299–315.

[7] Feng, L., Dillon, T., & Liu, J. (2001). *Inter-Transactional Association Rules for Multi-Dimensional Contexts for Prediction and Their Application to Studying Meterological Data. Elsevier Science Publishers B. V.*, Vol. 37, pp. 85–115.

[8] Gruteser, M., & Grunwald, D. (2003). Anonymous usage of location-based services through spatial and temporal cloaking. In *Proceedings of the 1st International Conference on Mobile Systems, Applications, and Services*, DBLP, San Francisco, pp. 31–42.

[9] Kalnis, P., Ghinita, G., Mouratidis, K., & Papadias, D. (2007). Preventing location-based identity inference in anonymous spatial queries. *IEEE Transactions on Knowledge & Data Engineering*, 19, 1719–1733.

[10] Shao, F., Cheng, R., & Zhang, F. G. (2014). A full privacy-preserving scheme for location-based services. In *Proceedings of the 33rd Information and Communication Technology — EurAsia Conference*, Springer, Berlin, Heidelberg, Bali, pp. 596–601.

[11] Montazeri, Z., Houmansadr, A., & Pishro-Nik, H. (2016). Achieving perfect location privacy in Markov models using anonymization. In *Proceedings of the 40th International Symposium on Information Theory and ITS Applications*, Shibata, pp. 355–359.

[12] Lu, H., Jensen, C. S., & Man, L. Y. (2008). PAD: Privacy-area aware, dummy-based location privacy in mobile services. In *Proceedings of the 7th ACM International Workshop on Data Engineering for Wireless and Mobile Access*, DBLP, Vancouver, pp. 16–23.

[13] Liu, X. X., Liu, K. K., Guo, L. K., Li, X. L., & Fang, Y. G. (2013). A game-theoretic approach for achieving K-anonymity in location based services. In *Proceedings of the 32nd INFOCOM*, Torino, pp. 2985–2993.

[14] Niu, B., Li, Q. H., Zhu, X. Y., Gao, G. H., & Li, H. (2014). Achieving K-anonymity in privacy-aware location-based services. In *Proceedings of the 33rd IEEE INFOCOM*, Toronto, pp. 754–762.

[15] Zhang, H. L., Xu, Z. K., Zhou, Z. G., Shi, J. T., & Du, X. J. (2015). CLPP: Context-aware location privacy protection for location-based social network. In *Proceedings of IEEE International Conference on Communications*, London, pp. 1164–1169.

[16] Li, H. X., Zhu, H. J., Du, S. G., Liang, X. H., & Shen, X. M. (2016). Privacy leakage of location sharing in mobile social networks: Attacks and defense. *IEEE Transactions on Dependable & Secure Computing*, 1, 1–14.

[17] Song, D., Sim, J., Park, K., & Song, M. (2015). A privacy-preserving continuous location monitoring system for location-based services. *International Journal of Distributed Sensor Networks*, 2015, 14.

[18] Pingley, A., Zhang, N., Fu, X. W., Choi, H. A., Subramaniam, S., & Zhao, W. (2011). Protection of query privacy for continuous location based services. In *Proceedings of the 30th IEEE INFOCOM*, Shanghai, pp. 1710–1718.

[19] Ghinita, G., Kalnis, P., Khoshgozaran, A., Shahabi, C., & Tan, K. L. (2008). Private queries in location based services: Anonymizers are not necessary. In *Proceedings of SIGMOD*, Vancouver, pp. 121–132.

[20] Shao, J., Lu, R. X., & Lin, X. D. (2014). FINE: A fine-grained privacy-preserving location-based service framework for mobile devices. In *Proceedings of the 33rd IEEE INFOCOM*, Toronto, pp. 244–252.

[21] Zhong, P. X., & Lu, R. X. (2015). PAD: Privacy-preserving data dissemination in mobile social networks. In *Proceedings of the 14th IEEE International Conference on Communication Systems*, Macau, pp. 243–247.

[22] Dwork, C. (2006). Differential privacy. *Lecture Notes in Computer Science*, 26, 1–12.

[23] Shokri, R., Theodorakopoulos, G., Troncoso, C., Hubaux, J. P., & Boudec, J. Y. L. (2012). Protecting location privacy: Optimal strategy against localization attacks. *CCS*, 617–627.

[24] Kellaris, G., Papadopoulos, S., & Papadias, D. (2017). Engineering methods for differentially private histograms: Efficiency beyond utility. *IEEE Transactions on Knowledge and Data Engineering*, 31(2), 315–328.

[25] Aggarwal, C. C., Sun, Z., & Yu, P. S. (2015). Fast algorithms for online generation of profile association rules. *IEEE Transactions on Knowledge & Data Engineering*, 14, 1017–1028.

[26] Sahoo, J., Das, A. K., & Goswami, A. (2015). An effective association rule mining scheme using a new generic basis. *Knowledge & Information Systems*, 43, 127–156.

Index

CPSIA information can be obtained
at www.ICGtesting.com
Printed in the USA
BVHW061946130720
583507BV00001B/2